EQUIPPING TOMORROW'S MILITARY FORCE

Integration of Commercial and Military Manufacturing in 2010 and Beyond

Committee on Integration of Commercial and Military Manufacturing in 2010 and Beyond

Board on Manufacturing and Engineering Design

Division on Engineering and Physical Sciences

National Research Council

NATIONAL ACADEMY PRESS
Washington, D.C.

National Academy Press • 2101 Constitution Avenue, N.W. • Washington, DC 20418

NOTICE: The project that is the subject of this report was approved by the Governing Board of the National Research Council, whose members are drawn from the councils of the National Academy of Sciences, the National Academy of Engineering, and the Institute of Medicine. The members of the committee responsible for the report were chosen for their special competences and with regard for appropriate balance.

This study was sponsored by the Joint Defense Manufacturing Technology Panel, U.S. Department of Defense, under contract No. N00014-96-D-0301. The views and conclusions contained in this document are those of the authors and should not be interpreted as necessarily representing the official policies, either expressed or implied, of the U.S. government.

International Standard Book Number 0-309-08316-8

Additional copies of this report are available from:

National Academy Press
2101 Constitution Avenue, N.W.
Box 285
Washington, DC 20055
800-624-6242
202-334-3313 (in the Washington metropolitan area)
http://www.nap.edu

Available in limited quantities from:

Board on Manufacturing and Engineering Design
2101 Constitution Avenue, N.W.
Washington, DC 20418
202-334-3124
bmaed@nas.edu

Cover: Image across the middle is a close-up of a circuit board.

Printed in the United States of America

THE NATIONAL ACADEMIES

National Academy of Sciences
National Academy of Engineering
Institute of Medicine
National Research Council

The **National Academy of Sciences** is a private, nonprofit, self-perpetuating society of distinguished scholars engaged in scientific and engineering research, dedicated to the furtherance of science and technology and to their use for the general welfare. Upon the authority of the charter granted to it by the Congress in 1863, the Academy has a mandate that requires it to advise the federal government on scientific and technical matters. Dr. Bruce M. Alberts is president of the National Academy of Sciences.

The **National Academy of Engineering** was established in 1964, under the charter of the National Academy of Sciences, as a parallel organization of outstanding engineers. It is autonomous in its administration and in the selection of its members, sharing with the National Academy of Sciences the responsibility for advising the federal government. The National Academy of Engineering also sponsors engineering programs aimed at meeting national needs, encourages education and research, and recognizes the superior achievements of engineers. Dr. Wm. A. Wulf is president of the National Academy of Engineering.

The **Institute of Medicine** was established in 1970 by the National Academy of Sciences to secure the services of eminent members of appropriate professions in the examination of policy matters pertaining to the health of the public. The Institute acts under the responsibility given to the National Academy of Sciences by its congressional charter to be an adviser to the federal government and, upon its own initiative, to identify issues of medical care, research, and education. Dr. Kenneth I. Shine is president of the Institute of Medicine.

The **National Research Council** was organized by the National Academy of Sciences in 1916 to associate the broad community of science and technology with the Academy's purposes of furthering knowledge and advising the federal government. Functioning in accordance with general policies determined by the Academy, the Council has become the principal operating agency of both the National Academy of Sciences and the National Academy of Engineering in providing services to the government, the public, and the scientific and engineering communities. The Council is administered jointly by both Academies and the Institute of Medicine. Dr. Bruce M. Alberts and Dr. Wm. A. Wulf are chairman and vice chairman, respectively, of the National Research Council.

COMMITTEE ON INTEGRATION OF COMMERCIAL AND MILITARY MANUFACTURING IN 2010 AND BEYOND

MICHAEL F. McGRATH, Sarnoff Corporation, Arlington, Virginia, *Chair*
WALTER B. LaBERGE, Naval Postgraduate School, Aptos, California, *Vice Chair*
ARDEN L. BEMENT, JR., Purdue University
RADM PETER DeMAYO, Lockheed Martin Corporation (retired), Alexandria, Virginia
GARY L. DENMAN, GRC International (retired), Pinehurst, North Carolina
JOSEPH A. HEIM, Genie Industries, Redmond, Washington
F. SUZANNE JENNICHES, Northrop Grumman Corporation, Linthicum, Maryland
JAMES MATTICE, Universal Technology Corporation, Dayton, Ohio
DAVID R. SMITH, Eastman Kodak Company, Rochester, New York
ROBERT I. WINNER, R. Winner & Associates, Hopkinton, Massachusetts

NRC Staff

PATRICK J. DOYLE, Program Officer

BOARD ON MANUFACTURING AND ENGINEERING DESIGN

JOSEPH G. WIRTH, Raychem Corporation (retired), Mt. Shasta, California, *Chair*
F. PETER BOER, Tiger Scientific, Inc., Boynton Beach, Florida
JOHN BOLLINGER, University of Wisconsin, Madison
PAMELA A. DREW, The Boeing Company, Seattle, Washington
ROBERT EAGAN, Sandia National Laboratories, Albuquerque, New Mexico
EDITH M. FLANIGEN, UOP Corporation (retired), White Plains, New York
JOHN W. GILLESPIE, JR., University of Delaware, Newark
JAMIE C. HSU, General Motors Corporation, Warren, Michigan
RICHARD L. KEGG, Milacron, Inc. (retired), Cincinnati, Ohio
JAY LEE, United Technologies Research Center, East Hartford, Connecticut
JAMES MATTICE, Universal Technology Corporation, Dayton, Ohio
CAROLYN W. MEYERS, North Carolina A&T University, Greensboro
JOE H. MIZE, Oklahoma State University (retired), Stillwater
FRIEDRICH B. PRINZ, Stanford University, Palo Alto, California
JAMES B. RICE, JR., Massachusetts Institute of Technology, Cambridge
JOHN B. STENBIT, TRW, Inc., Fairfax, Virginia
DALIBOR F. VRSALOVIC, AT&T Labs, Menlo Park, California
JOEL SAMUEL YUDKEN, AFL-CIO, Washington, D.C.

NRC Staff

TONI MARECHAUX, Director
ARUL MOZHI, Associate Director

Preface

As this report is being written, the United States and its allies find themselves attacked in a war of terrorism that vividly demonstrates the unpredictability and intensity of the threats that will challenge the credibility of U.S. national policy over the next decades. Although the specifics of this war were unforeseen, the Department of Defense (DOD) has been concerned for the past several years with the need for new kinds of flexible response capabilities and with the transformation needed for national security in a new global environment.

Each military service has been vigorously embarked on a technology-based transformation of its strategy, tactics, and force structure to significantly increase its ability for near-instantaneous projection of overwhelming military power to distant areas of U.S. national interest. This National Research Council study, *Equipping Tomorrow's Military Force: Integration of Commercial and Military Manufacturing in 2010 and Beyond*, deals with a co-requisite to this transition—namely, the need to also transform the U.S. military-industrial production support base.

Tapping into the strength of our commercial sector is a necessary element of any military strategy that relies on the ability to introduce new technology rapidly into operational use and on fast surge replenishment of weapons, spare parts, and other consumables vital to readiness and sustainability. It is also the only affordable path to keeping reasonably up to date with commercial technology advances.

This report was requested by the DOD Joint Defense Manufacturing Technology Panel to identify trends and opportunities for DOD and defense prime contractors to make optimal use of the technology and flexible manufacturing capabilities emerging within the U.S. commercial sector. The study committee appointed by the NRC Board on Manufacturing and Engineering Design consisted of members with backgrounds in defense manufacturing processes and operations, weapons system design, industrial engineering, and commercial manufacturing processes and operations.

In the course of its deliberations, the committee heard briefings from DOD personnel who participate in demonstrations of the integration of commercial and military manufacturing (ICMM), DOD acquisition managers and policy makers, and representatives of commercial firms that successfully integrate military and commercial business. The committee also reviewed numerous earlier reports on ICMM, a number of which are referenced in this report. The committee considered these reports and briefings as examples of successful experiences that can become part of the normal course of business in DOD.

The committee believes that a great opportunity exists for DOD to capitalize on the availability in the next decade of flexible, highly automated commercial and dual-use production lines for subsystems and components. Commercial supply chains will routinely exchange Web-based engineering and business data that drive efficient automated parts and assembly processes. Much of the commercial technology will be useful, directly or with customization, for advanced weapons systems applications. Weapon systems integration will still be the responsibility of defense prime contractors in 2010. They will increasingly have to consider new trade-offs involving commercial parts, commercial tooling and practices, design for commercial manufacturing, and commercial product support and upgrades. However, these trade-offs will make optimal use of the commercial base if, and only if, today's barriers and disincentives are removed.

What needs to be done to increase ICMM is well documented in earlier studies and in the examples cited in Chapter 3. One of the contributions this study hopes to make is to focus the attention of DOD decision makers and prime contractors on *how* to motivate and institutionalize needed implementation.

Michael F. McGrath, *Chair*
Walter B. LaBerge, *Vice Chair*
Committee on Integration of Commercial and
Military Manufacturing in 2010 and Beyond

Acknowledgments

The Committee on Integration of Commercial and Military Manufacturing in 2010 and Beyond would like to thank the following individuals for their presentations: Brench Boden, Wright-Patterson AFB; Jim Fallon, M/A-COM; Jacques Gansler, University of Maryland; Larry Griffin, consultant; LTG Paul Kern, U.S. Army; Steve Linder, Office of Naval Research; Fenner Milton, U.S. Army Night Vision Center; Spiros Pallas, Office of the Secretary of Defense; Lt. Col. Walter Price, U.S. Air Force; Herm Reininga, Rockwell-Collins; Stan Soloway, Deputy Under Secretary of Defense, Acquisition Reform; Jerry Thomas, SCI Systems; Joe Thomas, M/A-COM; Larry Trowel, General Electric Aircraft Engines; and Steve Werner, SCI Systems. This committee would also like to thank Leo Plonsky, Office of Naval Research, for his assistance during the study process.

This report has been reviewed by individuals chosen for their diverse perspectives and technical expertise, in accordance with procedures approved by the National Research Council's Report Review Committee. The purpose of this independent review is to provide candid and critical comments that will assist the authors and the institution in making the published report as sound as possible and to ensure that the report meets institutional standards for objectivity, evidence, and responsiveness to the study charge. The review comments and draft manuscript remain confidential to protect the integrity of the deliberative process. We wish to the thank the following individuals for their participation in the review of this report:

Rodica A. Baranescu, Navistar International Transportation Corporation,
Richard L. Kegg, Milacron, Inc. (retired),

Michael Lippitz, consultant,
Noel Longuemare, Principal Deputy Under Secretary of Defense, Acquisition and Technology (retired),
James B. Rice, Jr., Massachusetts Institute of Technology,
Phyllis Scales-Brown, U.S. Army Soldier and Biological Chemical Command (SBCCOM),
Anne Marie Velosa, Beaverton, Oregon, and
Jack White, Altarum Institute.

Although the reviewers listed above have provided many constructive comments and suggestions, they were not asked to endorse the conclusions or recommendations, nor did they see the final draft of the report before its release. The review of this report was overseen by James Solberg, Purdue University. Appointed by the National Research Council, he was responsible for making certain that an independent examination of this report was carried out in accordance with institutional procedures and that all review comments were carefully considered. Responsibility for the final content of this report rests entirely with the authoring committee and the institution.

Finally, the committee gratefully acknowledges the support of the staff of the Board on Manufacturing and Engineering Design, including Patrick J. Doyle, Program Officer; Toni Maréchaux, Director, Board on Manufacturing and Engineering Design; and Arul Mozhi, Associate Director.

Contents

EXECUTIVE SUMMARY 1

1 INTRODUCTION 6
Statement of Task, 7
Study Methodology, 7
Definition and Scope, 8

2 A COMPELLING CASE FOR THE INTEGRATION OF COMMERCIAL AND MILITARY MANUFACTURING 10

3 CURRENT SUCCESSES IN THE INTEGRATION OF COMMERCIAL AND MILITARY MANUFACTURING 14

4 OPPORTUNITIES FOR FUTURE INTEGRATION 22
A Spectrum of Choices for Manufacturing Integration, 23
Integration Opportunities Enabled by Technology, 23
Commercial Product Technology Trends, 24

5 BARRIERS TO INCREASED INTEGRATION 30

6 EDUCATION AND TRAINING 36
Understanding the Commercial Marketplace, 37
A Commercial Acquisition Academy, 38
Attracting Commercial Suppliers, 39
Research and Technology for Commercial Buying, 39

7 FINDINGS AND RECOMMENDATIONS 40
Findings, 41
Recommendations, 44
Discussion of Recommendations, 49

8 REFERENCES 52

APPENDICES

A BIOGRAPHICAL SKETCHES OF COMMITTEE MEMBERS 57
B BRIEFINGS 61
C BACKGROUND AND HISTORY 63
D ACRONYMS AND ABBREVIATIONS 68
E MANTECH STATUTE 70

Executive Summary

INTEGRATION OF COMMERCIAL AND MILITARY MANUFACTURING

More Than Just Commercial Off-the-Shelf

The integration of commercial and military manufacturing (ICMM) has been a subject of extensive debate and steadily increasing policy implementation in recent years. This integration can be defined as optimal use of the commercial manufacturing base to meet defense needs over the life cycle of a system. It encompasses a range of approaches, with commercial off-the-shelf (COTS) items at one extreme and products and processes unique to defense applications on the other. The intermediate approaches (shaded areas in Figure ES-1) are more complex, where defense needs are met with enhanced commercial products or by military products built in commercial or dual-use facilities.

The framework of Table ES-1 applies at the system, subsystem or assembly, and component levels. Commercial suppliers are willing to sell COTS products to the Depatrtment of Defense (DOD) using commercial practices supported by current policies. At the component level (particularly for electronics), defense contractors are increasingly using commercial components. Assemblies and subsystems, however, are predominantly built on military-unique production lines. The opportunity for 2010 and beyond lies in increasing the use of the commercial production base at the assembly and subsystem levels. These opportunities exist across mechanical, optical, and electronic subsystems.

Component

Subsystem

System

	COTS	Dual-Use Manufacturing			Defense-Unique
Product	Commercial	Enhanced Commercial	Unique	Unique	Unique
Manufacturing Process	Commercial	Commercial	Commercial	Enhanced Commercial	Unique

FIGURE ES-1 Commercial-military integration framework.

A Compelling Military Need

The expected nature of combat in 2010 is driving DOD to a substantial transformation of its force structure. The new force will be more flexible, able to respond more rapidly, and better equipped to deal with an unpredictable threat. The transformation strategy implicitly relies on the rapid introduction of new technology and rapid industrial response for the replenishment of weapons, spare parts, and other consumables essential to readiness and sustainability. The fact that potential adversaries have easy access to commercial technology will compel DOD and defense contractors to excel at being the first to integrate militarily relevant commercial technology into defense systems. An effective and robust integration of commercial and military manufacturing can improve military acquisition capabilities and capacity dramatically. It is needed in both long-cycle processes—those that produce technological marvels whose need and costs have to be predicted 15-20 years in advance—and processes that respond rapidly to changing threats and make the latest technology available to the warfighter at substantially lower unit and life-cycle costs.

Opportunities Abound

This study reviewed several dozen examples of successful integration of commercial and military manufacturing, ranging from pilot projects to sustained initiatives across many types of products. A few salient examples are shown in Table ES-1. Military and commercial and dual-use manufacturers have demonstrated the ability to work together in many different ways using many business models. In almost all of these examples, however, an individual on the govern-

TABLE ES-1 Examples of Successes in ICMM

Project	Results
DARPA/USAF miniature air-launched decoy (MALD)	Pushed the envelope to demonstrate that a high-performance, military-unique system could be developed from commercial sources
Motorola Communications Systems Division—JSTARS common ground station	Subcontracted 80% of circuit board production to commercial sources and realized higher quality and lower costs
Litton Amecom—NASA satellite control system	Used commercial standards, parts, and design practices to achieve cost reductions of 20-50%
USAF ManTech/TRW Military Parts from Commercial Lines Pilot Program	Demonstrated that electronic modules for the F-16 and RAH-66 could be redesigned for commercial manufacture and produced on a high-volume commercial line, with savings of 30-50% in cost and 90% in time from design release to production
USAF Electronic Systems Command-North—warning system unattended radar (AN/FPS-24)	Signal processor (ground environment) used commercial plastic encapsulated microcircuits to reduce costs by 85% and increase reliability (more than 6 million hours without failure)
USAF ManTech C-17 horizontal stabilizer outer torque box	Combined military and commercial production and achieved a 50% cost avoidance
M/A-COM Northrop Grumman ALQ-135 EW System	Commercial redesign for technology insertion reduced costs by 52%

ment side, and often one on the contractor side, had to make heroic efforts to use existing law and policy creatively to overcome barriers in common practice and culture. These barriers include the following:

- The lack of a commercial knowledge base and the need for an acquisition workforce (including contracting officers) familiar with commercial practice;
- Government acquisition provisions that commercial suppliers are unwilling to accept, including cost accounting, auditing, specialized specs and standards, procurement laws and socioeconomic provisions, intellectual property rights provisions, and logistics practices;
- Acquisition and upgrade cycle times that are drastically misaligned with commercial cycles, including lengthy test and evaluation and requalification requirements;

- Profit policies that discourage defense prime contractors from commercial outsourcing and push make/buy decisions toward "make"; and
- Lack of institutionalized solutions that program managers can routinely use.

In addition to the encouragement provided by previous successes, several trends will increase the opportunities for integration of commercial and military manufacturing over the coming decade:

- The emergence of a new $70 billion industry sector devoted to flexible electronics manufacturing services with capabilities suited to small-lot custom manufacturing for defense applications involving digital, analog, and microelectromechanical subsystems (MEMS);
- Advances in precision automated manufacturing for small lots, with greatly improved flexibility to accommodate on-demand manufacturing of defense items;
- Availability of technology for automated design, production, and in-line inspection and testing that, by 2010, will offer to electromechanical and MEMS devices a level of manufacturing flexibility and integration comparable to the production infrastructure for today's chips and boards;
- Commercial upgrades to aging avionics, a major opportunity for ICMM because upgrading using mil-spec configurations is usually unaffordable and often impossible; and
- Advances in the ability to share digital product data and business data in supply chains, with new efficiencies in rapid response manufacturing of the type needed by DOD.

RECOMMENDATIONS

Recommendations That Would Motivate Change

- Vigorously pursue policy, incentives, and implementing guidelines for ICMM. Such policies would create metrics and incentives for ICMM and remove barriers. An example of an appropriate metric would be to assess commercial content based on bills of material. Some potential incentives include the following:
  —Using source selection factors to assess trade-offs between commercial and military manufacturing,
  —Instituting profit policies to make defense contractor use of commercial production a profitable and viable business decision, and
  —Sharing any savings attributable to ICMM.

- Contract for life-cycle support and technology refreshment. Long-term contracts will create incentives for contractors to reduce costs, increase availability, and keep up with technology.

Recommendations That Would Enable Change

- Establish a commercial acquisition academy to augment training and education. The academy would have a research component to develop best practices and case studies as well as a training component to incorporate commercial practices into military acquisition training and processes.
- Fund and execute additional rapid-response demonstration programs to build a broad base of experience with ICMM. One obvious mechanism for this is the DOD ManTech program. Such demonstration programs could be an integral part of research programs in a commercial acquisition academy.
- Increase DOD and defense sector awareness of planned and emerging commercial technologies and capabilities. Market awareness is critical and can be easier for the military as a unique customer than for commercial competitors.
- Invest in research and development to increase the compatibility of military operating environments and commercially produced components.

1

Introduction

Owing to the expected nature of combat in 2010, U.S. military forces face a pressing need to transform themselves for rapid response to an unpredictable threat. Since DOD cannot be equipped for all contingencies, a necessary part of the transformation strategy will be an industrial base capable of satisfying the following needs:

- Flexible, fast response to military requirements for technological superiority given the situation at hand and
- Rapid replenishment of military consumables essential to readiness and sustainability.

Today's defense acquisition system and industrial base processes impose unacceptable delays in meeting both needs. By contrast, the global commercial industrial base thrives on rapid introduction of new technologies and on fast, flexible supply chains. Rapid advances in commercial technology (particularly in electronics), coupled with the easy access to commercial technology enjoyed by potential adversaries, will compel DOD and defense contractors to excel at integrating commercial technology into defense systems.

In many instances, this will require specifying and acquiring military variants of commercial components and subsystems produced on the same lines (or at least in the same ways) as commercial products. This integration of commercial and military manufacturing (ICMM) has begun on a small scale. By 2010, it needs to increase substantially if U.S. forces are to retain a technological edge.

Statement of Task

The overall objective of this study is to assess the opportunities for increased integration of commercial and military manufacturing in 2010 and beyond. Specific tasks are outlined below:

- Identify advances in commercial technology (e.g., flexible tooling, flexible manufacturing, and agile manufacturing) and best practices that will present opportunities for increasing the use of commercial enterprises in the manufacture of defense products in the next decade.
- Identify technology areas where rapid commercial advances in the next decade could be leveraged to improve military applications (e.g., computing, telecommunications, and biomedical technologies).
- Identify major weapons systems suitable for production by commercial enterprises in the next decade.
- Identify barriers limiting integration of commercial and military manufacturing (e.g., availability of spare parts, commercial willingness to produce defense products, design and tooling compatibility, and security issues) and methods for overcoming these barriers.
- Recommend strategies for optimizing the integration of commercial and military manufacturing in the next decade (e.g., developing the processes and technologies needed to increase the production of military products by commercial enterprises), including specific recommendations for DOD's Manufacturing Technology (ManTech) Program.

Box 1-1

STATEMENT OF TASK

This NRC report was requested by the DOD Joint Defense Manufacturing Technology Panel. Its objectives were to assess the opportunities for increased ICMM in 2010 and beyond, identify barriers, and recommend strategies for overcoming them. Specific tasks are shown in Box 1-1.

STUDY METHODOLOGY

The NRC Board on Manufacturing and Engineering Design appointed a study committee with backgrounds in defense manufacturing processes and operations, weapons system design, industrial engineering, and commercial manufacturing processes and operations (see Appendix A for biographical sketches of the committee members). In the course of its deliberations, the committee heard briefings from DOD personnel who participate in demonstrations of ICMM, DOD acquisition managers and policy makers, and representatives of commercial firms that successfully integrate military and commercial business. Briefings to the committee are listed in Appendix B.

The committee also reviewed numerous earlier reports on ICMM and decided to rely on these reports and briefings rather than create new case data on the current state of ICMM. Chapter 3 of this report summarizes examples of successful ICMM experiences and benefits. Chapter 4 identifies barriers to ICMM, many of which were identified in the earlier studies, which often recommended policy changes to deal with these barriers. Many of these policy changes were made in the course of acquisition reforms over the past 10 years. Appendix C summarizes the earlier studies, their recommendations, and policy changes. Appendix D lists the acronyms and abbreviations used in this study.

In the view of the committee, the current barriers to ICMM call not for more policy changes but rather for policy implementation, incentives, and culture change. These are the focus for recommendations in Chapter 7 of this report.

DEFINITION AND SCOPE

For the purpose of this study, the integration of commercial and military manufacturing (ICMM) means *optimal use of the commercial production base to meet defense needs over the system life cycle.*

- *Integration* is used to mean a complementary combination of commercial and military capabilities in manufacturing supply chains or within a manufacturing facility.
- *Commercial* implies a manufacturing base whose business goals, technical capabilities, and business practices are driven by commercial markets rather than government acquisition practices.
- *Military* implies both defense-unique capabilities unavailable in the commercial sector that must be maintained and defense-specific capabilities that may or may not be defense-unique.
- *Manufacturing* is used in the broad sense of a complete manufacturing enterprise, including design, fabrication, and support activities at all tiers in the supply chain.

ICMM is not limited to the procurement of commercial off-the-shelf (COTS) end items for defense use. Such a narrow view would miss the broad range of opportunities for commercial sourcing of components and subsystems that can be integrated into higher-level systems by defense prime contractors. The opportunities include both COTS components and defense-unique parts and subsystems built on commercial or dual-use (producing both military and commercial products) lines. Because straightforward procurement of COTS end items is improving in the defense procurement system, this study focuses on the more complicated issues involved in dual-use manufacturing and in COTS insertion in weapon systems.

To do business with leading commercial firms, DOD and defense contractors will need to understand and implement the best commercial practices. These practices include (1) integrated product and process development, (2) commercial design, production planning, and decision processes, and (3) assuring that manufacturing processes and technologies are proven before entering production. These practices are critical to assuring that defense components and subsystems can be manufactured in commercial facilities, and they must be adopted completely, not piecemeal, and with total commitment. Reliance on commercial business practices rather than on government acquisition regulations is essential. DOD should use defense-specific manufacturing only when it results in unique capabilities that cannot be found or developed in the commercial sector.

Primarily because of the large and increasing importance of systems integration, there will be little opportunity for major weapons systems to be produced as a whole by commercial enterprises over the next decade. The committee expects the greatest impact of ICMM to be on subsystems and components.

Global Sourcing

Global sourcing is an increasingly important part of modern commercial manufacturing, but the committee considered it to be beyond the scope of this study. The full military implications of global sourcing of components and subsystems are difficult to assess. In areas where national security risks are introduced by dependence on foreign sources, defense manufacturing may need mitigation strategies beyond those adopted by commercial manufacturers. The military faced this issue before for commercial integrated circuits and solved it by building up parts inventories to prevent military shortages. However, as ICMM graduates from individual components toward higher-level assemblies from commercial lines, inventory costs like this might wipe out the savings from ICMM (the committee has not done the economic modeling to come to a definite conclusion). Additional strategies will be needed to address this issue.

Protection may also be needed to prevent unauthorized design features that can act as trapdoors or Trojan horses, although this can be an issue with domestic sources as well, even defense-specific sources.

In light of the complexity of these issues, the implications of foreign sourcing might be an appropriate topic for a separate study. However, the fundamental findings and recommendations for increasing ICMM presented here are appropriate regardless of decisions related to global sourcing.

2

A Compelling Case for the Integration of Commercial and Military Manufacturing

Several factors point to a compelling need for DOD to adopt ICMM as a routine way of doing business for the production and life-cycle support of DOD weapons system components, assemblies, and subsystems. Key factors are the importance of sustaining technological superiority, a declining production base for new weapons, and advantageous trends in the commercial manufacturing sector.

Today, military decision makers are unsure of who our future enemies will be, as well as when, where, and under what environments and rules of engagement future conflicts will take place. To respond to these uncertainties, a flexible and responsive industrial base will be critical. Our military decision makers today are trapped in an acquisition strategy that depends on an industrial base that cannot respond quickly enough to meet the demand for new and modified military systems expected to result from the stepped-up tempo of future military operations. Rapid and reliable access to the products and services of the commercial industrial base and use of commercial practices could make acquisition more responsive to the current environment. A robust implementation of ICMM can change our military acquisition capability from one of long cycle times producing technological marvels whose need and costs have to be predicted 15 to 20 years in advance to a capability that flexibly responds to changing threats.

Technological superiority underpins our military strategy. Today, however, the commercial sector leads the military industrial sector in developing and adopting new technology, particularly for electronics components. The overall lead of the commercial sector is illustrated in Figure 2-1.

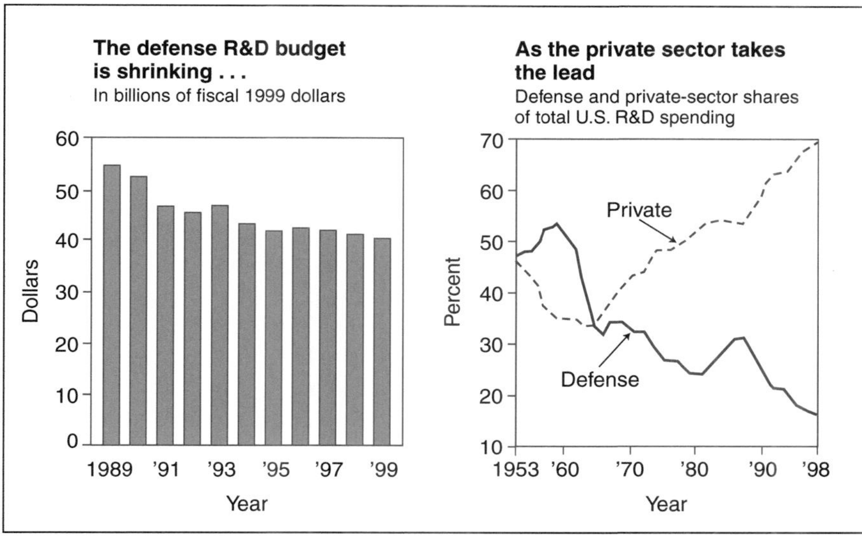

FIGURE 2-1 Defense vs. private R&D funding. From USD(AT&L) (2001), which assembled data from many sources. The sources here are the U.S. Office of Management and Budget and the National Science Foundation.

Furthermore, our current and potential adversaries have ready access to most commercially available technology-based products, and it is impossible to control the export and diffusion of critical new technologies that feed emerging global commercial markets. In the long range, for example, such new technologies as microelectromechanical devices (MEMS) will emerge from the global commercial industry, and by 2010 the military must be ready to exploit the power of this type of technological advancement ahead of our adversaries. A recent Defense Science Board report on the technology capabilities of non-DOD providers (DOD, 2000a) recognizes the need for DOD to find and acquire commercial technology and to increase its acquisition of technical assistance from sources outside DOD and traditional defense contractors. In limited cases, export controls can slow the movement of these technologies outside our borders, but they cannot stop it.

One clear solution to this military parity issue is for DOD to have direct access to commercial technologies and to adopt them much more quickly than our adversaries. The committee believes that adoption of ICMM as an industrial base strategy can significantly increase the pace of technology insertion and help to sustain technology superiority. Just as the military has become almost totally dependent on the commercial sector for microcircuits, its greater use in future commercial integrated electronic subsystems does not threaten the stability of the military industrial complex. Rather, ICMM is a winning strategy for the military if we can master the issues associated with reducing the lead time to introduce new and updated systems.

The potential for ICMM to benefit life-cycle support of military systems is equally compelling. Military budget constraints through 2010 will severely limit the number and production rate of new platforms (aircraft, ground systems, and ships). Aging military systems and the obsolescence of subsystems and parts will be increasingly common. Subsystem upgrades will solve some of the obsolescence problems and provide new military capabilities. However, even with a robust upgrade program, without ICMM it is unlikely that the military can keep pace with the electronics technology being introduced by the commercial sector on a new product cycle as short as 9 months but not more than 3 years. The commercial industry, driven by competitive pressures, successfully deals with this modernization issue every day, taking a total value perspective. Frequently, commercial products are modernized by either rapid introduction of new models or insertion of product upgrades that are compatible in form, fit, and function. For example, an advanced disk drive for a microcomputer can now be inserted in a plug-and-play environment. In many cases, military systems can be designed in this way. One effort to promote this design approach is the modular open systems architecture (MOSA), promoted by DOD's Open Systems Joint Task Force.

A commercial modernization strategy can be adopted by the military by assigning the defense prime contractors total product responsibility, including product support. The contractors and their commercial subcontractors would be responsible for replacing a product (subsystem) if it fails. During replacement, the contractor could choose to modernize components, if improved parts are available, on a form-fit-function basis. This strategy has been successfully implemented on a few military systems and subsystems (see Chapter 3), and wider adoption of this life-cycle support concept through the use of the commercial manufacturing base can give the military reliable, high-quality systems for the entire life cycle at lower cost.

Some segments of commercial industry are changing to take advantage of contract manufacturing. This is a growing industrial sector, particularly in electronics manufacturing, where the pace of technology advances is rapid and the interdependence of product technologies and manufacturing processes is high (Miscioscia and Libin, 2000). Contract electronics manufacturing is a high-growth sector that is expanding from a build-to-print model to a capability that integrates product design, manufacturing, and product support. DOD contractors are beginning to use these specialized electronics manufacturers, but only to a very limited extent.

A single business model obviously does not fit all of the situations the committee studied. Examples of commercial manufacturing of military electronics include Rockwell-Collins, where a single process concept has been implemented for design and production of both military and commercial electronic systems in common facilities;[1] M/A-COM, which produces both military and commercial microwave integrated circuits using common processes and facilities;[2] and SCI, a commercial electronic manufacturing service (EMS) company that produces large volumes of commercial electronics assemblies and systems as well as some military products using common processes.[3]

Expanded use could be made of contract manufacturers as a new source of competition based on a new model for military electronics design and production. Also, DOD could significantly benefit from access to short-production-cycle capabilities inherent in these specialized manufacturers, the continuing technology refreshment demanded by the commercial marketplace, and the benefits of commercial-scale learning curves that provide reliable, high-quality, competitively priced products.

Finally, expanded adoption of ICMM can lead to significant cost savings. Several examples of implemented ICMM, particularly for electronic systems, clearly demonstrate the potential for very impressive cost savings, sometimes of 50 percent or more. While the examples are not extensive, the cost savings analyses are credible and of such magnitude that profit-on-profit[4] concerns related to the subcontracted work are not significant. While cost savings and the resulting affordability of new systems provide a strong case for ICMM, the more compelling arguments are those regarding technology superiority and access to a flexible and responsive industrial base.

In summary, it is essential that our military be equipped with the latest technology and that the technology be superior to that of our potential adversaries. Further, the unique systems integration capability of the U.S. defense industry can be enhanced through aggressive adoption of ICMM, which will be well worth the sustained commitment of management this adoption will require.

Some military-unique products and technologies cannot be produced by the commercial sector, and a few should not be, due to the need for assured access and security. These exceptions are usually developed through DOD R&D activities, and the ability to acquire these products and technologies while preventing their dissemination to our adversaries must be maintained.

---

[1]Herm Reininga, Rockwell-Collins, Dual Production Experience at Rockwell-Collins, presentation to the committee on January 26, 2001.

[2]Jim Fallon and Joe Thomas, M/A-COM, A Common Manufacturing Base for Dual-Use Applications, presentation to the committee on January 25, 2001.

[3]Steve Werner and Jerry Thomas, SCI Systems, Circuit Board Manufacturing, presentation to the committee on January 25, 2001.

[4]Profit-on-profit refers to a prime contractor earning a percentage of profit on profit earned by a subcontractor.

3

Current Successes in the Integration of Commercial and Military Manufacturing

Several programs have successfully integrated commercial and military manufacturing, supporting the essential finding that ICMM can succeed and provides worthwhile benefits. Successful ICMM implementations exist in electronics, aircraft engines, unmanned vehicle structures and guidance systems, and others, but they represent only a small fraction of expenditures.

Some of the cases the committee studied were demonstration efforts (e.g., MALD[1] and TRW MPCL[2]). Others were ongoing production programs and long-term initiatives (e.g., Rockwell-Collins,[3] GEAE,[4] and SCI Systems[5]). Table 3-1 summarizes the results the various cases achieved. Winner and Griffin (1998) detail all the cases in this table, unless otherwise specified.

Many significant lessons and conclusions can be drawn from the cases the committee studied. The first of these is that opportunities are increasing for the defense industry to outsource manufacturing services for both electronic and

[1]LTC Walter Price, USAF, Wright-Patterson Air Force Base, Miniature Air-Launched Decoy (MALD) Program, presentation to the committee on January 25, 2001.

[2]Brench Boden, Wright-Patterson Air Force Base, Military Products from Commercial Lines (MPCL), presentation to the committee on April 12, 2001.

[3]Herm Reininga, Rockwell-Collins, Dual Production Experience at Rockwell-Collins, presentation to the committee on January 26, 2001.

[4]Laurence M. Trowel, General Electric Aircraft Engines, Military and Commercial Turbine Engines, presentation to the committee on January 25, 2001.

[5]Steve Werner and Jerry Thomas, SCI Systems, Circuit Board Manufacturing at SCI Systems, presentation to the committee on January 25, 2001.

structural/mechanical assemblies. The commercial world recognizes the economic and competitive benefits of outsourcing manufacturing to facilities capitalized to operate at larger scale. For example, a large new sector has emerged: flexible electronics manufacturing services (EMS) companies. These EMS companies—for example, SCI Systems, which is mentioned in Table 3-1—are expanding from manufacturing services into engineering design for large commercial companies. Security clearances to permit access to classified data and designs are obtainable where necessary, and some manufacturers are willing to get them and abide by security rules.

TABLE 3-1 Benefits of ICMM

Performer—Project	Actions (Summary)	Results
DARPA/USAF—miniature air-launched decoy (MALD), overall[a]	Used existing systems and subsystems; CAIV fly-away target cost; COTS and GOTS equipment; commercial practices and standards, low-cost manufacturing approach; design for flexibility; nontraditional production industry	Reduced payload material costs 72%; achieved average unit cost target of $30,000; performance better than predicted
DARPA/USAF—miniature air-launched decoy (MALD), airframe[a]	Automobile manufacturing process for airframe; commercial sheet molding compound materials	Reduced development, tooling, recurring costs; average unit cost $4,500; simplified assembly; light weight; vibration dampening
DARPA/USAF miniature air-launched decoy (MALD), engine[a]	Used nontraditional manufacturers for engine; design for reduced touch labor; catalog parts	Reduced from 121 detailed parts in traditionally designed engine to 33; 4 hours touch labor
General Electric Aircraft Engines (GEAE)—COTS engines[b]	Commercial engines used in military applications	Costs reduced, e.g., in KC135R, E-3, KC-10, E-4, Air Force One, ABL, C-5
GEAE—J85 Propulsion Modernization Program[b]	Using commercial design components from CJ610/CF700: stator casting, spooled rotor, cast mainframe, combustion liner assembly	Longer life, lower unit and maintenance costs

Table continued on next page

TABLE 3-1 Continued

Performer—Project	Actions (Summary)	Results
GEAE—shared commercial and military practices[b]	TF39, TF34, J85, LM2500 determined commercial; government source inspection eliminated; engines on performance-based payments, spares on delivery-based payment; engines and spares catalogs on price analysis	In 2000, more than 40 percent of military sales on commercial basis, with more planned 21 MIL-STDs replaced with GEAE processes
USAF—Arnold Engineering Development Center[b]	Military facilities used for commercial turbine engine testing; repeatable processes; automated test facilities	Very high fraction of U.S. commercial and military turbine engine altitude tests performed in integrated facilities
Motorola Communications Systems Division—overall	Subcontracts 80% of 1997 circuit board production to commercial suppliers (vs. 5% in 1985)	Higher quality, lower costs, QA workforce reduced 60-70%
Motorola Communications Systems Division—JSTARS common ground station	Military parts and production transition to COTS parts and subsystems, commercial production	Cost reduced from $6 million to about $1.4 million
Litton Amecom—NASA satellite control system	Interactive design process involving customers; commercial bus-interface standards used; design for use of commercial space parts	Cost reduced 25% from predecessor; next generation reduction projected to be 33% less than current (for total of 50% reduction)
Rockwell Collins—single process initiatives	Rated and nonrated components purchased together	50% reduction in purchase orders
Rockwell Collins—single process initiatives	Evaluation of all military and civilian processes	All product-flow processes very near or same as previous commercial processes; operating costs reduced $30 million in first year; another $30 million anticipated for second year; flexible dynamic assignment of personnel across plants; workload balancing raised utilization rates to over 90%

TABLE 3-1 Continued

Performer—Project	Actions (Summary)	Results
Rockwell Collins—single process initiatives	Reinvestment of savings	Tracking state-of-the-art process technologies while maintaining competitive prices
Rockwell Collins—Army precision lightweight GPS receiver (PLGR)	Single process; commercial parts; commercial process; maintenance and availability are contractor's responsibility	Direct labor content, 3-4% (rivals best commercial, high-volume rate); eliminated parts obsolescence problem
Rockwell Collins—Navy ARC-210 radio	Redesign for commercial parts and processes; annual design reviews; maintenance and availability responsibility with contractor	Price reduced 42%; field reliability (MTBF) increased from 500 flight hours to 807 (+62%); annual cost reduction and reliability improvement; eliminated parts obsolescence
Rockwell Collins—USAF Pacer CRAG	Design for commercial parts and processes; dual-use	Savings (est.) of $90 million on award of about $235 million
USAF/TRW—military products from commercial lines (MPCL) program—CIM	Developed flexible high-volume commercial and low-volume military CIM and improved design-manufacturing interface	Design release to production reduced labor 90% (200 hours to less than 20)
USAF/TRW MPCL—radio frequency front-end controller (RF/FEC) modules for F-22 and RAH-66 CNI	Redesign for civilian (automotive division) production	73% cost reduction ($40,000 to $10,800)
USAF/TRW MPCL—pulse narrow-band preprocessor (PNP)	Redesign for civilian (automotive division) production	54% cost reduction ($34,000 to $15,500)
USAF/TRW MPCL—general	Civil-military manufacturing process integration and quality improvement via design of experiments	11 of 14 (so far) critical processes at $C_{pk} > 1.33$; PWA assembly 100% automated
USAF/TRW MPCL—new business model	Developed, documented detailed business model acceptable to government and many commercial contractors	Survey of 1,340 companies estimated 68% cost savings from military baseline for sample modules

Table continued on next page

TABLE 3-1 Continued

Performer—Project	Actions (Summary)	Results
U.S. Army ManTech/CPI, Northrop Grumman, Teledyne—radars	Military and civilian radar dual production line cost reduction; $1.6 million invested	$19 million PAC-3 cost avoidance
U.S. Army ManTech—PEM protection	Adapted commercial IC chips for harsh environment and long-term storage; $5.8 million invested	$357 million cost avoidance over six aviation and missile systems
USAF ManTech—C-17 horizontal stabilizer outer torque box[c]	Combined commercial and military production; reduced government oversight and reporting	Greater than 50% cost avoidance
USAF Electronic Systems Center (ESC)—generic PWA manufacturing process	Integrated civil and military processes via evaluation of critical control elements for each process step	Of 156 military and commercial specs and standards, 49 retained, 20 recommended for replacement, 87 not needed
USAF Electronic Systems Center (ESC)/Rockwell-Collins—test of above PWA process	Compared military-specific, commercial, and new dual-use process using military and commercial components on digital control board from GRC-171 radio	25-35% reduction in assembly, labor, and overhead costs for dual-production and commercial processes; additional 20-30% cost reduction predicted feasible with commercial flat-pack components; dual-production and commercial processes equal to or better than quality of military process; production throughput increased approx. 30%; no process-related failures in accelerated thermal test (17 lifetimes)
USAF ESC—North warning system unattended radar (AN/FPS-24) signal processor (ground environment)	Substituted commercial plastic encapsulated microcircuits (PEM) for military parts	No failures in >6 million operating hours (2/1995); still operating (8/1998) with no commercial PEM failures; 87% average cost reduction for microcircuits (range = 85-92%)

TABLE 3-1 Continued

Performer—Project	Actions (Summary)	Results
USAF ESC/Rockwell-Collins—JTIDS receiver/ synthesizer PWA (uninhabited fighter environment)	Evaluated military certified parts, commercial parts meeting military temperature requirements, and commercial parts using both military and commercial assembly	Military assembly costs 15% higher; using mil-temp passive and 10 commercial (of 33) active components reduced material cost by 23%; quality as good as or better than for commercial process; commercial component failure rate lower than military; process-related failures equal, even with more inspection in military process
USAF ESC/GEC-Marconi—PLSR RF module	Commercial parts and processes substituted for military	65% reduction in material cost, 30% reduction in labor for assembly, test; same electrical performance
Boeing/SCI Systems—Longbow Apache helicopter computer and intercom[d]	Built-in dual production facilities using commercial and military parts	1997-2002 scheduled delivery: 2,800+ computers and 707 intercom units
SCI Systems—catalog parts[d]	Dual production and design builds catalog of standard products	Used in AAR47 microprocessor, joint tactical terminal, flight simulators, radar-controlled chassis, B52 pylon tester computer, Navy databus couplers, and Firefinder radar processor
USN/SCI/Eaton—Virginia class nuclear propulsion controls[d]	Military subsystems built in dual-use facilities	Backplanes, module assemblies, circuit card assemblies built in SCI's integrated facility
M/A-COM—common RF microwave and millimeter wave technology manufacturing base[e]	Designed boards, assemblies, products, and manufacturing processes for dual use	Produce in dual-use facilities: GaAs MMIC amplifiers, ball-grid arrays for base station switching and phased-array radars, radar front-ends, radio circuit boards, microwave cable assemblies, millimeter-wave transceivers, voice over IP radio systems, etc.

Table continued on next page

TABLE 3-1 Continued

Performer—Project	Actions (Summary)	Results
M/A-COM Northrop Grumman—ALQ-135[e]	Redesign for technology insertion, best-value manufacturing, IPPD, strategic partnership to address commercial and military markets	52% cost reduction
Navy/Lockheed Martin—FBM Program (TRIDENT)[f]	Open system architecture—COTS insertion to address technology, obsolescence, supportability, and cost issues	Increased COTS products as a percent of parts to 60%, attained a 75% parts count reduction, a 50% development cycle time reduction, and a cost avoidance of $1.2 billion

NOTE: The source of all information for unfootnoted projects is Winner and Griffin (1998).

[a] LTC W. Price, U.S. Air Force, Miniature Air Launched Decoy (MALD) Program, presentation to the committee on January 25, 2001.
[b] L. Trowel, General Electric Aircraft Engines, Military and Commercial Turbine Engines, presentation to the committee on January 25, 2001.
[c] S. Linder, Office of Naval Research, Overview of ManTech Program and Sponsor's Goals, presentation to the committee on October 13, 2000.
[d] S. Werner and J. Thomas, SCI Systems, Circuit Board Manufacturing at SCI Systems, presentation to the committee on January 25, 2001.
[e] J. Fallon and J. Thomas, M/A-COM, A Common Manufacturing Base for Dual-Use Applications, presentation to the committee on January 25, 2001.
[f] Lockheed Martin (2001).

A second observation is that ICMM promotes the exploitation of emerging technologies and enables their rapid delivery to the customer; however, buyers, users, and industry must cooperate closely to accomplish this. This is exemplified by almost all of the cases in Table 3-1.

As noted above, several of the cases the committee studied were demonstrations, with the associated benefits and issues. Demonstration programs, pilots, prototypes, and the like are useful to discover what is possible, prove out new approaches and processes, and learn lessons. However, almost every case required extraordinary contracting actions. For example, in the USAF/TRW MPCL case, a contracting officer had to make an unprecedented determination that the circuit boards in question were "commercial parts" in order for TRW automotive to be willing to manufacture them on its commercial lines. The DOD needs to capture the procedures used in these demonstrations and develop guidelines or standard approaches for similar activities. Demonstration projects should include the development of results and how-to documents; their dissemination is critical but not currently well done.

The committee observed the important point that demonstration programs are often underfunded, causing them to take years longer than needed. They should be funded for 2-year completion, at most.

Commercial and military design processes differ substantially, because design objectives and schedules differ. The demonstrations illustrate that to make ICMM work, integrated product and process development (IPPD) can and must be used to make sure that military designs are producible at reasonable cost in commercial facilities. For example, in electronics, commercial and dual-use capacity exists, as does the willingness to make ICMM work. Low-lot-size, high-mix facilities are available, but site-specific design rules must be followed from concept on, an example of IPPD. One substantial problem in both electronics and other product types is that many defense designs unnecessarily preclude commercial manufacturing by specifying tolerances and traceability requirements more stringent than those used in commercial processes rather than relying on design for robustness and mature production processes (Winner and Griffin, 1998; GAO, 1998).

Another lesson is that appropriate incentives—such as in the ARC-210 case in Table 3-1—can achieve substantial success for both the military customer and the industrial supplier. In the ARC-210 program, the contractor has configuration control within the system as long as it meets availability and form, fit, and function interface requirements. A multiyear descending price curve was contracted in advance, and the contractor has maintenance responsibility as well as all the incentive necessary to decrease the cost and increase the reliability of the system as rapidly as possible.

From the listed cases, the committee members' experience, and some informal enquiries in the industry, the committee observed a trend toward defense contractors increasing their outsourcing of electronic board manufacturing (both fabrication and populating) and their use of commercial components. (See, for example, Motorola and SCI in Table 3-1.) The same kinds of forces that took defense industries out of the integrated circuit business, especially the capital cost of keeping up with technology and competitiveness, lead to outsourcing at higher levels of assembly. The military acquisition commands can and should accelerate this trend.

Some of the programs reported in the table were ManTech projects. ManTech is chartered by Congress and enjoys unique authorities that can be used to address aspects of a transition to greater use of ICMM (see Appendix E). However, the program is not adequately funded or tasked to implement further significant ICMM demonstrations. Early successes highlight the potential for ManTech to play a pivotal role, through demonstration efforts in technology, business practices, and partnerships, in realizing a more efficient ICMM industrial base.

4

Opportunities for Future Integration

ICMM will increase in importance as a means for DOD to develop and enhance the nation's defense technological superiority. Product and production process technology trends that are observable now will, by 2010, make ICMM even more attractive to DOD than it is today as a means of dealing with the significant challenges that will shape the defense acquisition climate, including the following:

- System complexity will increase, driven by rapid advances in technology, the growing technological sophistication of potential adversaries, and a higher dependence on technology because of the changing nature of warfare—for example, unmanned, network-centric warfare, monitored and coordinated from space, and global asymmetric threats.
- Defense budgets will continue to be under pressure, making it difficult to deliver more complex systems with the limited funds available. Advanced R&D investment in defense will compete with investments increasingly required to modernize, upgrade, and retrofit aging defense systems.
- DOD is likely to experience a serious shortage of scientists and engineers qualified to conduct advanced R&D. This conclusion is based on current workforce demographics and declining numbers of graduate students in important fields of physical sciences and engineering, including physics, chemistry, metallurgical and materials engineering, and aeronautical engineering and increased competition from cutting-edge commercial industry (NRC, 2001).

Given these constraints and challenges, DOD will need more options for tapping into the strength of the commercial industrial base.

A SPECTRUM OF CHOICES FOR MANUFACTURING INTEGRATION

The tails of the ICMM spectrum are clear: unique defense solutions on one end and COTS on the other. Defense manufacturing is evolving. At one time, unique solutions were the norm, but DOD is now facing the reality that such solutions are costly, that acquisition time is excessive, and that pressure on defense spending, as well as technological and business trends, necessitates more affordable approaches. The opportunities for increased ICMM dictate considering the choices between the extremes of COTS and defense-unique manufacturing.

The committee developed a commercial-military integration framework (Figure 4-1) that identifies ICMM opportunities between unique solutions and COTS. This framework provides a structure for the options available for ICMM that the committee found useful for developing strategy and recommendations.

The basis for this ICMM framework is to consider both the product being purchased, i.e., the defense component or system, and the manufacturing process to produce it. COTS represents commercial products made by commercial processes, while the opposite end of the framework represents unique defense products made by unique defense processes. The opportunities between these end points include a range of dual-use manufacturing options:

- Enhanced commercial products manufactured on commercial production lines,
- Commercial products manufactured on enhanced commercial production lines,
- Unique defense systems manufactured on commercial production lines, and
- Unique defense systems manufactured on enhanced commercial production lines.

"Enhanced" implies some improvement in product design or process capability so that the commercial source better meets defense (or dual) needs. The ICMM framework and ICMM options form the basis for the committee's strategy recommendations.

INTEGRATION OPPORTUNITIES ENABLED BY TECHNOLOGY

As DOD scans the R&D horizons and industry's technology roadmaps for

Component

Subsystem

System

	COTS	Dual-Use Manufacturing			Defense-Unique
Product	Commercial	Enhanced Commercial	Unique	Unique	Unique
Manufacturing Process	Commercial	Commercial	Commercial	Enhanced Commercial	Unique

FIGURE 4-1 Commercial-military integration framework.

leverage opportunities, a broad view of ICMM suggests that the answers to the following questions may serve as leading indicators of opportunity:

- In which applications will technology trends deliver commercial *products* that fully satisfy defense requirements, with at most a modest enhancement to the product design or the commercial production process? It would be important for DOD to identify such opportunities early to favorably influence its military systems and permit the fielding of new technologies by the United States before they are available to the nation's adversaries.
- In which applications will technology trends deliver commercial production *processes* capable of manufacturing defense-unique designs? Again, it would be important for DOD to identify early those emerging production technologies where no more than a modest enhancement would be needed to fully satisfy defense requirements. It would be equally important to identify where unique defense systems could be designed to fit the manufacturability constraints of emerging production processes.

COMMERCIAL PRODUCT TECHNOLOGY TRENDS

The technology trends discussed here represent future possibilities by the year 2010 but are by no means an all-inclusive forecast of the advanced commercial technologies that could be leveraged by DOD over the next decade. In general, the commercial product trends that will be most relevant to DOD as ICMM opportunities will be in areas where the commercial product application is similar to that of the defense product and DOD can obtain low-volume access to high-volume production for a variant of the product, enhanced (if necessary) for military use. (Other ICMM opportunities will arise from process trends, rather than products, where commercial manufacturing can be configured to run defense product on the same line.)

The economics of high-volume markets enables commercial industry to advance technology and reduce costs at a pace that defense could never afford. This has been and continues to be the case in microelectronics. It is also the case for Global Positioning System (GPS) technology, cellular phones, and turbine engines. High-volume commercial production enables a faster descent of the learning curve and a firmer statistical foundation for in-line process control, six-sigma quality, and higher process yields of reliable parts.

Some product technology trends to consider as future ICMM opportunities include the following:

- *Wireless communications technology.* At the component level, GaAs components and assemblies for cell phones and military use are coproduced, and the trend will continue as new wide-bandgap materials are developed. Higher-level assemblies with advanced analog-digital conversion, signal processors, and steerable phased-array antennas will also offer ICMM opportunities.
- *Photonics.* The trend toward integration of photonics and electronics at the chip level will open possibilities for increased ICMM, especially in communications systems. Commercial high-powered lasers will also be candidates for military applications in targeting, imaging, and countermeasures.
- *Night vision systems.* Uncooled infrared (IR) imagers are now entering the commercial market and will dwarf the military market by 2010. Low-cost IR devices can revolutionize night vision, missile seekers, and other applications. This commercial technology can also give potential adversaries capabilities that they could not previously afford.
- *Biomedical technology.* The rapid advances in human genomics, microfluidic chips, computational drug discovery, and medical instrumentation will have direct COTS application to military healthcare as well as ICMM potential for biowarfare defense products.
- *Fuel cells.* Commercial developments in fuel cells, from palm-size to industrial-size generators, offer the military an important means for powering the electronics and stealthy vehicles of 2010 and beyond.
- *Microprocessors.* Over the next decade, microprocessors will be economically viable as substitutes in the functions currently delivered by integrated circuits (ICs), application-specific integrated circuits (ASICs), and discrete electronic components. Electronic functions will be software

reconfigured to the task at hand. This trend will reduce electronics costs, increase system reliability by reducing the number of parts, and have the net effect of reducing the number of unique electronics subsystems required in defense systems.

- *Micro-opto-electromechanical systems (MOEMS).* This technology trend offers such great promise as an ICMM opportunity that the committee singled it out for further discussion in the section below.

Microscale Systems Combining Optical, Electronic, and Mechanical Devices

For several years, MEMS technology (or MOEMS, if optics is included) has been viewed as an important technology with the potential to affect the 21st century as much as microelectronics technology affected the 20th century. In fact, among the unique attributes of MEMS or MOEMS is the opportunity to achieve low cost through semiconductor-like batch processing on wafers, where thousands of devices can be produced essentially at the cost of producing just one (Tang, 1999).

MEMS is an enabling technology that can be leveraged across broad product markets. This technology represents a unique opportunity for ICMM in multiple applications, including information and communications, health and life sciences, automotive, aerospace, measurement and control, and power systems.

The timing is right to accomplish ICMM through MEMS technology. MEMS is an emerging technology, and MEMS factories (also called "fabs," just as semiconductor chip factories are called) of the future are emerging as well. With appropriate planning and investment, creating MEMS fabs capable of producing defense and commercial devices on the same production line will be straightforward.

In the defense world, DARPA is already leading the way in exploiting MEMS technology for military applications. For example, DARPA has already developed a MEMS-based platform stabilization system that replaces $1,000 worth of conventional accelerometers and gyroscopes with a single chip that costs $20 (Global Information, Inc., undated).

Given the priority and R&D investment being applied to MEMS technology in Europe and Asia, it is not clear that the United States will emerge as the global leader, as it did in microelectronics (Detlefs, 1998). Embracing MEMS as an ICMM opportunity, to the extent that it drives R&D investment in MEMS applications as well as the processes to manufacture them, could have significant implications for future U.S. global competitiveness.

Commercial Manufacturing Processes and Business Trends

The committee noted several macro trends in commercial design, manufacturing, and business processes that will create a climate for increased ICMM opportunities:

- *E-commerce.* Tools and methods for e-commerce that can drive automated manufacturing in supply chains are advancing rapidly in the commercial world. They will become a ubiquitous and important enabler of ICMM well before 2010, an easy port of entry to commercial manufacturing for DOD.

  DOD and defense prime contractors will be able to access and understand the process capabilities of commercial suppliers rapidly, and DOD will be able to issue solicitations and access competitive quotes online, globally if necessary, with fast cycle times. With a clear understanding of process capabilities and competitive costs across the commercial marketplace, DOD will be well positioned to make appropriate sourcing decisions.

  Electronic commerce will have important implications for DOD in the future. For example, with the existence of digital product databases, make-to-order spare parts could greatly reduce defense inventories and their associated costs. In addition, the e-commerce capabilities discussed above can be used to rapidly increase the total production capacity for military goods in case of a crisis by using industry-accepted methods of data exchange to enhance communications with new suppliers.
- *Contract manufacturing.* Contract manufacturing is becoming standard practice in commercial industry for circuit boards, for plastic molding, and even for the manufacturing and testing of subsystems or final products. Contract manufacturing, as described in Chapters 2 and 3, offers DOD and its prime contractors a significant ICMM opportunity by leveraging a flexible, high-volume manufacturing base with multibillion-dollar assets, thereby achieving economies of scale in purchasing and production. Use of this production base will permit application of broad, high-volume experience to defense manufacturing, and the resulting learning curves will lead to lower cost and higher quality.
- *Precision engineering.* Consumer products, particularly consumer electronics, are becoming more and more miniaturized, fueled to a large degree by product designers in Asia. Miniaturization here leads to products that are more portable or that incorporate more features in the same package. The trend toward miniaturization drives smaller physical dimensions and proportionately smaller dimensional tolerances. These trends require commercial manufacturing processes and tooling capable of higher precision than in the past. As commercial manufacturing evolves to higher

precision capability, opportunities will open up for DOD to leverage this capability for defense precision manufacturing needs that were once beyond the scope of commercial production.

- *Structural and functional integration.* Wherever possible, commercial designers are making use of embedded microprocessors and integrated analog circuits, sensors, and actuators to reduce the number of parts and assembly operations. Clever design approaches to multifunctional integration can achieve not only substantial cost reductions but also improved performance. In addition, parts rationalization and a reduction in the number of fasteners are often achieved by the use of complex precision castings, forgings, and sheet stampings that integrate structural functions and employ integral fastening methods.
- *Advanced CAD/CAM.* By 2010 it will be possible to go routinely from "art to part" in a totally automated process; that is, it will be possible to create the three-dimensional solid model of a mechanical part, transfer the model electronically to the supplier of choice, and automatically program a "fixtureless" numerically controlled (NC) machine to make the part with essentially no human intervention. Further commercial advances could make it possible to go well beyond automated NC machining. It is possible that the three-dimensional solid model could also drive the automated flexible assembly of parts and automated calibration and testing of assembled products, although it is not clear how far commercial investment will drive this technology. Ultimately, with appropriate investment, the end result would be true mass customization in which the manufacture of a single item becomes viable. Rapid prototyping has already advanced sufficiently to realize this goal for some classes of parts.

Despite the broad success of U.S. companies during the 1990s, there is a steady retreat from domestic manufacturing in many fields (NRC, 1999b). This may add risk to ICMM. While contract manufacturing is a significant trend, few contract manufacturers of circuit boards are based in the United States. DOD will have to assess the risks of depending on foreign suppliers to enable ICMM and take appropriate action.

Importance of Early Visibility

Commercial markets will continue to drive technology advances where volumes are high and profitability potential justifies the technology investments. Microelectronics was a classic example of this in the 20th century. Where the commercial technology advances result in COTS solutions, leverage by DOD is straightforward. Otherwise, commercial technology advancements will proceed but leverage by DOD will be difficult. Given the competitive pressures on commercial manufacturers today, defense needs are unlikely to be of interest, particularly if defense volumes are low.

Therefore, DOD should engage early with applicable commercial technology trends, particularly where product specs or process capability must be tweaked to accommodate a defense need, so that the technology advancement opportunity can be seized.

Special Defense Requirements

Commercial technology advances are generally targeted for use in environments that are far less hostile than many defense systems must withstand. Environmental compatibility can prevent the adoption of commercial technology advances for defense needs and was always a common reason for specifying military-unique components in the past. The report *Defense Manufacturing in 2010 and Beyond* (NRC, 1999a) identifies the qualification of commercial parts for these harsh environments as a gap that needs to be addressed to allow making use of commercial manufacturing capacity. These environmental considerations include zero gravity, vacuum, radiation, temperature and humidity extremes and cycles, vibration and shock, and electromagnetic interference.

The *Defense Manufacturing* report also states that DOD will have to qualify commercial parts that are not specifically designed to withstand these environments and modify them to meet military needs or develop system designs to compensate for their limitations. Where harsh environmental conditions are a serious requirement for a defense application of commercial technology, early involvement in development will be critical to assure that new commercial components can withstand or be adapted to withstand military operating conditions.

5

Barriers to Increased Integration

The desire to use more commercial products and processes in defense systems has been a significant goal of DOD for many years. Many earlier studies underscored this need, which is motivated by concerns about system acquisition and affordability, the dynamics of a rapidly consolidating or shrinking defense industrial base, and real worries over diminishing sources of defense-unique processes and products. It was not until 1994, however, that Secretary of Defense William Perry launched initiatives to reform the acquisition process.

In these new acquisition reforms, actions needed to simultaneously enable abolition of military-unique specifications and standards, encourage far greater use of COTS items, and support the evolution of a common civil-military industrial base became viewed almost interchangeably, without recognition of the different steps needed to implement these reforms. Much confusion was generated at the many policy implementation levels of DOD and the military services, especially because the reform efforts coincided with the beginning of post-Cold War workforce downsizing. In addition, reforms involving the rewriting of Federal Acquisition Regulations (FARs), legislative reform to relieve burdensome oversight, new contracting practices, and identification of pilot programs to test reforms were all proceeding simultaneously. Despite the progress in issuing policies in these areas, the goal of an integrated civil-military industrial base that would routinely yield defense systems with measurably greater (and more cost-effective) commercial content has remained elusive.

Sufficient time has now passed to permit planning and action to implement new ICMM policies. Industry consolidation is largely complete, at least among prime contractors, and workforce downsizing, while possibly incomplete, has reached a modicum of stability. Many individual examples of ICMM potential have been demonstrated, but they remain the exception rather than the rule (usually requiring acquisition professionals to take unprecedented actions, with some career risk). The revisions to the FAR provide potentially enabling guidance and authorizations, although mixed interpretations are still likely across different procurement activities.

Today, increased ICMM seems much more achievable than in the early days of acquisition reform, but there are still longstanding barriers that must be dealt with. Significant among these barriers are the following:

- *Requirements process.* Acquisition reform has made substantial advances in emphasizing performance-based requirements—expressing the warfighter's need in terms of function and military utility instead of specifying how to manufacture the product. However, the cycle time for validating needs and contracting for defense product development, production, and sustainment remains excessive. Opportunities for more aggressive product/process ICMM are lost because this cycle is grossly out of sync with the commercial life cycle of development, production, and system support. Further, while "cost as an independent variable" (CAIV) requirements have been embraced as an acquisition reform philosophy, cost targets have not been sufficiently aggressive or enforceable to motivate serious pursuit of ICMM trade-off alternatives.
- *Definition of commercial products and services.* Acquisition reform also made significant strides in developing a new definition of a commercial product or service in FAR 2.101. However, that redefinition is quite narrow, particularly with respect to enabling the procurement of R&D as a commercial activity. This is true even when the actual work closely parallels or seeks to tap into existing commercial R&D as a forerunner to what could become an ICMM-based acquisition. The definition problem is fundamental. The present situation results from a definition philosophy that seeks primarily to limit, rather than optimize, the ability to leverage the commercial marketplace where such leverage is greatest—in the R&D phase. Notably, in the R&D stage commercial industry (not yet having captured the market) is usually still willing to consider defense needs. This will be so only if the defense customer does not compromise rapid product deployment to the commercial marketplace, jeopardize intellectual property rights, or impede efficiency by using parallel, defense-unique production lines.
- *Incomplete use of FAR Part 12.* The Federal Acquisition Streamlining Act (1994) called for the executive branch to place heavy emphasis on the acquisition of commercial products and services to the maximum practical extent, so a new section was added to the FAR. FAR Part 12, Acquisition of Commercial Items, gives wide latitude to contracting officers to

buy commercially. Despite some successes, DOD has yet to realize the full potential of buying under FAR Part 12. Part of the reason is cultural and part is a lack of tools (e.g., market research and pricing tools) that the contracting workforce could use to buy commercially. These issues are addressed in Chapter 6. An aggressive and expanded implementation of FAR Part 12 is essential to the success of ICMM. In an ideal acquisition system, there would be sufficient flexibility and transparency to the type of technology used that it would be unnecessary to designate a product or service as commercial to reduce regulatory oversight. However, in the absence of such a system, more frequent designation of products or services as commercial offers significant benefit.

- *Profit policy*. This is a longstanding barrier in doing business with the DOD. It has been the subject of numerous studies in recent years, including a recent Defense Science Board Study (DOD, 2000b). The Under Secretary of Defense (Acquisition, Technology and Logistics) has as one of his five goals an initiative to improve the health of the defense industrial base. Profits on DOD contracts are largely governed by the provisions of the weighted guidelines in Department of Defense FAR Supplement (DFARS) Section 215-404-4.[1] These are detailed procedures that consider three factors in coming up with an overall profit objective: (1) performance risk, (2) contract type risk, and (3) level of capital investment in facilities. DOD contracting officers are well trained in calculating and applying the weighted guidelines in a way that limits the profits on DOD contracts. This practice of negotiating profit has no counterpart in the commercial sector. The committee examined the DOD profit policy provisions of the DFARS and considers that they are a significant barrier to achieving an aggressive implementation of ICMM. The primary reason is the emphasis the weighted guidelines put on facilities capital. Specifically, the provisions state that contractors are "encouraged and rewarded for aggressive capital investment in facilities" (DFARS 215-404-71-4). This incentive has long-term consequences because it encourages contractors to bring work in-house. It also frustrates an initiative like ICMM, the success of which depends on creative and innovative outsourcing, primarily to commercial suppliers. DOD acquisition executives should reexamine profit policy with an eye to dramatically stimulating outsourcing, particularly to commercial suppliers, by removing disparities in allowable profit rates on contracts for in-house vs. subcontracted work. This would lead to less costly systems. In the long run, it would cause prime contractors to reduce excess capacity and overhead (the extensive facilities and manpower to perpetuate military-specific custom design and

[1]The DFARS is available online at <http://web2.deskbook.osd.mil/htmlfiles/DBY_ dfars.asp>.

production of many items). In December 2000, DOD added a technology incentive to its weighted guidelines (DFARS 215.404-71-2). This incentive adds as much as 4 percent profit for technology innovation. This could be a fruitful area, particularly in the commercial sector. This new incentive has not received widespread implementation in DOD buying commands or in the defense industry, reinforcing the committee's impression that cultural barriers remain a major impediment to ICMM.

- *Absence of certain other incentives.* Other incentives should be considered by DOD to enhance ICMM. Longer-term contracts would motivate industry to make investments that yield savings over the product life cycle. Award term contracts could be used for this purpose. DOD could institute programs that give industry a share of the savings that accrue to DOD as a result of investments and other actions by industry. Value engineering is another tool that could be adapted to enhance ICMM and satisfy the needs of both government and industry.
- *Acquisition oversight.* It is noteworthy that most motivations for relief emanate from a desire to simplify the acquisition process, with the overseers dedicated to preventing abuse. The statutory warrants of both contracting and financial/audit management officials are brought to bear to prevent even the perception of wrongdoing, resulting in an acquisition environment between military customer and supplier that is largely adversarial. This is a totally different set of dynamics from that in the commercial customer-supplier world. It calls for a totally new basis for determining appropriate oversight—namely, oversight that does not unintentionally prevent the adoption of commercial products, processes, and practices, oversight that enables rather than impedes ICMM.
- *Cycle times.* Major defense weapon system research, development, testing, and evaluation (RDT&E) plus acquisition cycle times average 7 to 15 years, depending on the type of system, and have lifetimes of 20 to 50 years. By contrast, commercial product development cycles and product lifetimes are both shorter. The military system must be supported in the field with technology upgrades and maintenance/overhaul actions over an extended period that has few commercial counterparts. Yet the same product-life-cycle actions are accomplished in the commercial arena, including technology refreshment, only faster. This cycle time mismatch is a barrier that will probably never be completely resolved, but it certainly could be significantly improved. The new DOD 5000.1 policy of evolutionary acquisition could have an important impact by enabling technology upgrading using rapidly evolving commercial technology. Evolutionary acquisition will necessitate the definition of requirements incrementally over time.
- *Workforce knowledge and experience base in ICMM.* Modern defense acquisition professionals are highly trained, motivated, and focused on

meeting the needs of the warfighter. This workforce consists of functional specialists, including program managers, contracting officers, financial managers, engineers, logisticians, and other specialists. They recognize their role as statutory and regulatory stewards who protect the interests of the taxpayer. They have worked hard to become an integrated product team (IPT) that executes the acquisition function. However, the very depth of this specialized approach has inadvertently worked to frustrate the shift to robust ICMM. The culture, motivation, education, training, and reward system for this workforce has traditionally been driven by a defense-unique business and program world. It is not surprising that the few successful examples of ICMM, or even ICMM pilot demonstrations, have resulted from heroic efforts by some acquisition professionals. Rather than viewing ICMM as an extraordinary event, rarely to be duplicated, this workforce must become, over time, expert and steeped in commercial practice and in all aspects of commercial business. Such a transformation is fundamental to successful ICMM and will take a substantial investment of people and funding to be realized across the workforce. All aspects of acquisition professional development (from recruitment to experience patterns, education/training, motivation, promotion/assignment policy, and so on) will have to change. Once adopted, changes in contract incentives will tend to drive the commercial workforce in this direction. However, this experience base will have to be developed in both the contractor and the government workforces.

- *Low-risk demonstration.* The recognized successes of acquisition reform often stemmed from demonstration or pilot programs with special authorities to waive or modify current practices. In most cases, there was an experimental atmosphere, and program managers were not seen as risking their careers. In spite of this, real change (and there has been some) came slowly and was not easily transferred into general practice. DOD executives will need to execute a much more aggressive set of demonstration programs and authorities than prevailed in the initial reform movement, and they will need a higher level of education and training in commercial practices to make adoption of ICMM widespread. A much more robust level of experimentation is warranted, and significant additional tools and methods, such as modeling and simulation of technology and business practice variables in real and hypothetical acquisition scenarios, will have to be devised. Personnel will need to have broad expertise in industry business, design, and product support practices.
- *Lack of institutionalized solutions.* The program management teams of even the largest acquisition actions rarely have the autonomy to develop and implement ICMM solutions on a case-by-case basis in time to benefit their programs. Program managers and prime contractors need a climate of commercially oriented procedures that they can routinely draw upon

and a workforce steeped in commercial practice that can work from the outset of a program to increase ICMM.

Dealing with these longstanding barriers to the implementation of ICMM will be difficult, especially given their deep roots in the culture of defense acquisition. The progress to date in acquisition reform, a growing base of experience in ICMM, the trends in technology and commercial business practices cited in this report, and increased incentives for ICMM are all part of the solution.

6

Education and Training

As DOD seeks to move toward greater reliance on the commercial sector for its products and services, its acquisition workforce must be trained, because the success of ICMM will largely depend on a knowledgeable and competent workforce. This increased training effort requires the active participation of the Under Secretary of Defense (Acquisition, Technology and Logistics) (USD(AT&L)) and the service acquisition executives. This is particularly important because DOD's acquisition workforce will need to develop the requisite skills to deal effectively in the commercial marketplace. Training should fully support the acquisition process and allow DOD to take full advantage of the entire marketplace. Those involved in the delivery of a training program that emphasizes the commercial marketplace should be held accountable for results.

Hearings held on July 17, 2001, by the House Committee on Government Reform, Subcommittee on Technology and Procurement Policy, appear to support these efforts. In those hearings, the Director of Defense Procurement (Lee, 2001) acknowledged that DOD training needs should be updated to reflect commercial practices and the unique needs of the commercial sector. While the testimony addresses the specific needs of commercial intellectual property rights, it reflects a broader concern that the changed business environment will require flexibility in laws and regulations for dealing with the commercial sector. This testimony also notes DOD procurement policy makers have had a dialogue with firms that declined to do business with DOD because of intellectual property concerns.

UNDERSTANDING THE COMMERCIAL MARKETPLACE

Despite DOD's strong emphasis on the training and education of its military and civilian acquisition personnel, current training does not equip DOD personnel to understand the commercial marketplace.

DOD has a well-trained acquisition workforce with a high level of skill in the specialties of traditional defense acquisition, such as FAR-based procurement, engineering, test and evaluation, and logistics. However, it is not taking full advantage of commercial capabilities, because it does not provide training specialized in the commercial sector, most notably in commercial sourcing. It is well positioned to develop and implement a training program that would provide the knowledge and tools to take advantage of the technology, products, and services that exist in the commercial sector.

The Defense Acquisition University (DAU) has the management responsibility for all acquisition training within the DOD. DAU campuses are in place across the country. It has been a leader in the development and delivery of technology-based training through distance learning. The Defense Systems Management College (DSMC) is an important component of the DAU. Located in Ft. Belvoir, Virginia, the DSMC is responsible for courses leading to level III certification in acquisition management and for post-level III training for program executive officers (PEOs), program managers, and deputy program managers for major programs in terms of military priority and cost. These major programs are usually designated acquisition category (ACAT) 1 or 2, and the training requirements are rooted in law. The DAU trains in excess of 40,000 government personnel each year, and this level of training is expected to continue well into the future.

The Industrial College of the Armed Forces (ICAF), located at Ft. McNair in Washington, D.C., delivers a 1-year course for senior personnel who are expected to be future leaders in DOD and other government agencies. About one-third of the ICAF student body specializes in acquisition management. Keeping in touch with industry is an important part of the ICAF mission. Each of the students participates in an industry studies program that requires field trips to various private sector companies. This industry studies program can be an important resource for DOD as it seeks to expand its emphasis on ICMM and gain a greater understanding of the commercial industrial base.

Finally, DOD has recognized that it is facing a significant challenge posed by the demographic composition of its workforce, including those who work in acquisition. A substantial percentage of the DOD acquisition workforce is eligible to retire, and there are not enough people ready to replace them. In October 2000, DOD issued its final report of the Acquisition 2005 Task Force, "Shaping the Civilian Acquisition Workforce of the Future" (USD(AT&L), 2000), whose recommendations are now in the process of being implemented. The committee strongly endorses the thrust of that report. For the purposes of ICMM, its implementation represents an opportunity to bring new blood into the workforce and to develop in them skills that will allow DOD to take advantage of the commercial marketplace.

USD(AT&L) and the service acquisition executives need to put in place a comprehensive training program to develop expertise in the defense industrial base. The training program must recognize the changing business environment that exists for DOD and should be aimed at program managers and other acquisition personnel, including contracting officers. It should emphasize the technology as well as the processes, products, and services of the commercial sector. It should also give acquisition personnel new tools that they can use to buy better value for DOD and an improved capability for its operating forces. The essential elements of the training program are described next.

A COMMERCIAL ACQUISITION ACADEMY

The USD(A&T) could send a strong signal to the acquisition community, and create a strong agent for change, by establishing a commercial acquisition academy within the DAU organization to bolster training and education that emphasize the commercial marketplace. New curricula as yet unfamiliar to DOD will have to be developed and delivered by experts in the commercial sector, whose help and support DOD should enlist in designing a program that meets the needs of both. The curricula should provide tool-based knowledge so that acquisition personnel can buy effectively in the commercial sector under commercial rules while still protecting the interests of DOD and maintaining the necessary accountability to the taxpayer. Training should emphasize commercial best practices through case studies and other practical exercises. This will require integrating courses offered by the commercial acquisition academy with other DAU courses, including those at DSMC. Industry (including firms not doing military business) and business schools should be solicited to help in the development of case studies and to augment the faculty. The DAU also may be able to take advantage of courses in place at commercial companies. Industry associations can play a key role in curriculum development and facilitate participation by commercial companies.

All acquisition personnel should be trained in critical thinking, problem solving, and decision making as a prerequisite for level III certification. Post-level III training of PEOs and program management personnel should emphasize business acumen, critical thinking, and problem solving. Case studies, with particular emphasis on the commercial sector, should be the primary teaching tool. Special attention should be paid to benchmarking current commercial practices and using commercial terminology throughout the academy's programs. Differences in terminology between defense and commercial industry complicate the use of commercial suppliers and practices.

The curriculum should emphasize industry tours and internships, which let government personnel see how industry operates and bring good ideas back to DOD. Such tours and internships can also be a two-way street where industry people learn more about government. To the extent practicable, commercial personnel should teach DAU courses, particularly those offered at DSMC.

ATTRACTING COMMERCIAL SUPPLIERS

Program managers and contracting officers need specific training on how to overcome the barriers that discourage commercial firms from doing business with DOD. This training is extremely important and is likely to become more of an issue over the next several years, because DOD lacks capabilities that the commercial sector has to offer. A good first step would be broad-based training on what latitude exists in current law and regulation to procure products and services from commercial companies. This could then be augmented with targeted training on subjects like intellectual property and commercial pricing. This type of training needs to touch the whole workforce so that government deals with private industry on a consistent basis.

RESEARCH AND TECHNOLOGY FOR COMMERCIAL BUYING

The DOD Acquisition Research Program should be used to develop additional tools to facilitate commercial buying. This program, which is already in place, should be directed toward the commercial sector and address high-payoff areas such as intellectual property and commercial pricing. The results of acquisition research should be fed into the DAU curricula.

DOD needs to fully exploit ways of carrying out market research over the Internet and sharing lessons learned and best practices. The latter can by itself be an effective learning tool and become a way of doing business for all acquisition personnel.

In summary, training focused on commercial practices is essential to achieving meaningful progress in ICMM. While training is an essential part of the implementation of ICMM, leadership and constant management attention to the difficult process of cultural change, including attention to clear and implementable policies, metrics, and incentives, are necessary as well. This change requires a somewhat radical approach, like the proposed commercial acquisition academy, because DOD and its acquisition workforce do not understand the commercial sector well enough to realize how it can be used to improve the capability of DOD systems and processes.

7

Findings and Recommendations

Based on the analysis in the foregoing chapters, the committee developed the five findings and six recommendations discussed below. The recommendations are discussed relative to the specific points in the statement of task at the end of this chapter. The committee's recommendations are directed primarily at the DOD management chain above the DOD Manufacturing Technology Program (ManTech). However, it sees an important role for the ManTech program in the overall strategy for ICMM and has included specific recommendations for ManTech.

The committee's recommendations are as follows:

1. Vigorously pursue policy, incentives, and implementing guidelines for ICMM.
2. Establish a commercial acquisition academy to augment training and education.
3. Fund and execute additional rapid-response demonstration programs to build a broad base of experience with ICMM.
4. Contract for life-cycle support and technology refreshment.
5. Increase DOD and defense sector awareness of planned and emerging commercial technologies and capabilities.
6. Invest in R&D to increase the compatibility of military operating environments and commercially produced components.

The committee believes that bolstering training and education is the highest priority area. It recommends the establishment of a commercial acquisition academy within the Defense Acquisition University, the mission of which would be to create a knowledge base and an acquisition workforce steeped in commercial practice. The committee believes that this new institution could become the focal point for following through on the other findings and recommendations. Further details are presented in the section on recommendations, below.

FINDINGS

Finding 1. Defense systems integrators have the pivotal role in ICMM.

With rare exceptions, major defense systems require a combination of system integration skill and military application domain expertise that can come only from the established defense contractor community. For this reason, the greatest opportunity for ICMM exists at the component and subsystem level. System integration of superior technologies into defense systems, including defense-unique and commercial technologies, is the edge that the U.S. defense industry can provide. ICMM offers defense prime contractors the same rewards and business potentials as commercial original equipment manufacturers (OEMs), but only if DOD changes acquisition practices identified as barriers, especially profit policies and provisions that bias make/buy decisions.

Finding 2. Recent successes in ICMM show it can be done.

Although ICMM is not yet a routine practice, numerous experiments and pilot programs and a few real acquisitions have demonstrated its viability. Further efforts are needed to institutionalize successful demonstration results. When motivated by competition, cost constraints, contract incentives, or performance needs, defense prime contractors are willing to outsource the production of subsystems and components to commercial manufacturers. Experiments, pilot programs, and actual acquisitions show that commercial manufacturers are willing to accept government contracts and subcontracts if ways are found to work through the barriers created by acquisition practice. Successful ICMM implementations exist in electronics, aircraft engines, submarine electronics, and aging avionics, but they represent only a small fraction of expenditures. These implementations show that

- Appropriate incentives can lead to substantial successes.
- It is important that ICMM become part of the normal course of business and that it not require government contracting personnel to make unprecedented determinations that a product or service is commercial or to take other actions that put their careers at risk.

- The acquisition system must become more willing to embrace changes proven in demonstration programs and to disseminate results and how-to documents.
- The ManTech program, with its unique statutory authorities, can be used for demonstration efforts in technology, business practices, and partnerships to realize a more efficient ICMM defense industrial base. However, the program is not now tasked or funded for this mission.

Finding 3. Longstanding barriers must be removed to take full advantage of ICMM.

Many of the barriers to ICMM are acquisition and business practices rather than technical incompatibilities. Recent legislative and policy changes provide the needed tools, but implementation has been uneven across programs. Key barriers that continue to inhibit ICMM include the following:

- Lack of a commercial knowledge base and an acquisition workforce (including contracting officers) that is steeped in commercial practice;
- Government acquisition provisions that commercial suppliers are unwilling to accept, including cost accounting, auditing, specialized specs and standards, procurement laws and socioeconomic provisions, and logistics practices;
- Government practices on intellectual property rights that are incompatible with commercial practice;
- Acquisition and upgrade cycle times that are drastically misaligned with commercial cycles, including lengthy test and evaluation and requalification requirements;
- A requirements process for new systems that is often incompatible with evolutionary commercial capabilities and development practices and that does not set cost requirements based on ICMM opportunities;
- A definition of a commercial product or service (FAR 2.101) that significantly constrains the ability to procure commercial R&D;
- A lack of effort on the part of DOD commands to fully implement FAR Part 12, Acquisition of Commercial Items;
- Failure to take full advantage of DFARS profit policy provisions on technology innovation—continued emphasis on investment in new facilities acts as a disincentive to commercial outsourcing;
- Lack of modular equipment architectures, such as modular open systems architecture (MOSA) and of methods for scheduling and funding periodic block upgrades throughout the product life cycle; and
- Lack of institutionalized solutions that program managers can routinely draw on to make ICMM solutions part of the normal course of business in DOD.

Finding 4. Commercial trends make ICMM increasingly desirable in 2010 and beyond.

Commercial industry trends that make ICMM increasingly desirable and feasible for 2010 and beyond include more outsourcing, more automated and flexible production facilities, rapid changeover to manufacture a wide variety of subsystems, and efficient utilization of facilities that make a broad mix of products. The envelope of technical capability in commercial manufacturing facilities is generally large enough to allow production of defense variants of commercial products.

The committee finds that current DOD knowledge of commercial markets and products is inadequate for assessing emerging commercial products and capabilities and taking them into account in defense system plans and requirements. Defense contractors can help provide improved insight into the commercial base, but DOD needs its own (in-house or contracted) capability to be a smart buyer.

Among the significant trends that will open possibilities for increased ICMM are the following:

- The emergence of a new $70 billion industry sector devoted to flexible electronics manufacturing services, with capabilities suited to small-lot custom manufacturing for defense applications involving digital, analog, and microelectromechanical subsystems (MEMS);
- Advances in the mechanical manufacturing domain in precision automated manufacturing for small lots, with greatly improved flexibility to accommodate on-demand manufacturing of defense items;
- Availability of technology for automated design, production, and in-line inspection and testing that, by 2010, will offer electromechanical and MEMS devices with a level of manufacturing flexibility and integration comparable to the production infrastructure for today's chips and boards;
- Commercial upgrades to aging avionics, a major opportunity for ICMM, because upgrading using mil-spec configurations is usually unaffordable and often impossible; and
- Advances in the ability to share digital product data and business data in supply chains, with new efficiencies in rapid-response manufacturing of the type needed by DOD.

Finding 5. Current training does not equip DOD personnel to understand the commercial marketplace.

The DOD acquisition workforce, including program management staff and contracting officers, does not have the necessary background to take full advantage of commercial capabilities. While it is well trained and has a high level of skill in the specialties of traditional defense acquisition, this workforce lacks training on doing business in the commercial sector. Individual program managers cannot pioneer ICMM unless the supporting acquisition establishment understands how it works and actively supports its implementation. Training resources that can be leveraged for this purpose include the following:

- The Defense Acquisition University (DAU), DOD's primary source of training in acquisition, including the training of contracting officers. Annual throughput exceeds 40,000 students and is projected to remain at this level over the next several years.
- Private sector education and training, including government/industry personnel exchange programs.
- The Defense Systems Management College (DSMC), a component of DAU, the primary source of program management training. Training requirements for acquisition personnel are specified in law.
- DOD's Industrial College of the Armed Forces (ICAF), which—as part of its educational mission for DOD personnel—conducts studies to learn about industry practices and pass on the knowledge and experience.

RECOMMENDATIONS

Recommendation 1. Vigorously pursue policy, incentives, and implementing guidelines for ICMM.

The Under Secretary of Defense (AT&L) should issue specific policy statements and implementing guidelines that remove barriers to the use of commercial products and processes, that provide metrics for commercial content in military systems, and that monitor implementation in milestone reviews. Implementing guidelines should specifically include the following:

- *Source selection factors.* All source selections should include evaluation factors for performance and cost trade-offs that explicitly assess use of commercially produced subsystems. These factors must carry sufficient weight to serve as a strong incentive for industry to optimize the use of commercial technologies and manufacturing capabilities.
- *Profit policies.* Weighted guidelines for DOD profit policy should be revised and/or implemented to reduce the reward for facilities investment and increase the reward for technology innovation, including integration of commercial technology. If these reward provisions are properly structured, DOD prime contractors will be motivated to do more commercial outsourcing.[1] Changing DOD profit policy in this manner will require

[1]While issues such as those of legacy infrastructure complicate these decisions, they should not be permitted to impact the affordability of future systems.

the endorsement of senior leadership, an aggressive training program for acquisition personnel, and follow-up to see that changes are implemented.

- *Shared savings.* For programs beyond Milestone B, contracts should provide the contractor a generous share of any savings attributable to design or production changes that implement ICMM.
- *Intellectual property.* Authority should be granted for acquisition research demonstration programs to use "other transactions" authority until legislation is available to enable commercial intellectual property and data rights practices. Particular emphasis should be placed on the robust implementation of the technology innovation provisions of DFARS 215.404-71-2. In addition, DOD should seek changes in intellectual property and data rights laws to align DOD practice with commercial practice, including negotiation of field-of-use rights and limiting technology exchange to specific contracts, people, and durations using commercial nondisclosure agreements.
- *Commercial item determination.* Program managers and contracting officers should be given flexible guidance on how to determine if products or services are commercial items under FAR Part 12. In parallel, DOD should seek waivers and, ultimately, a change to broaden the FAR definition of a commercial item to include all items produced on commercial lines, and it should launch demonstration programs to deal with the challenge of pricing commercial items.
- *Metrics.* Such metrics as commercial content based on bills of material should be developed, implemented, and tracked in a variety of demonstrations in each service.
- *Request for Proposal (RFP) template.* An RFP template should be established for the DOD Acquisition Desk Book. This template should be aimed at fostering a shared understanding of commercial design standards and rules for government and industry representatives. In addition, government and prime contractor representatives should be encouraged to participate in industry bodies that set design and manufacturing standards.
- *Acquisition research.* DOD's Acquisition Research Program should be used to develop additional tools to facilitate buying from the commercial sector. Two suggested areas would be the development of commercial pricing techniques and the harmonization of DOD and industry intellectual property (IP) requirements.

Recommendation 2. Establish a commercial acquisition academy to augment training and education.

A comprehensive training program to develop expertise in the commercial industrial base, with particular emphasis on acquisition of commercial technology, products, and services, should be established under the sponsorship of the

USD(AT&L) and the service acquisition executives. This program should be directed at program managers and other acquisition personnel, including contracting officers. Creating an acquisition workforce steeped in commercial practice is of such importance to ICMM that a commercial acquisition academy should be established within the DAU and dedicated to this objective. A commercial acquisition academy should:

- Use case studies and practical exercises to impart commercial best practices. Industry and industry associations, both defense and commercial, as well as business schools should all participate in curriculum development (including case studies) and should make staff available to augment regular faculty. Where possible, industry personnel should attend DAU courses, particularly those at the DSMC.
- Offer coursework that helps program managers (PMs) and contracting officers identify latitude in the existing legal and regulatory language that would allow procurement of products and services from the commercial sector.
- Recruit faculty with skills in commercial industry practices, especially in the areas of commercial pricing techniques and commercial intellectual property issues. Industry associations, retirees, and commercial industry should be involved in faculty recruitment.
- Train both program managers and contracting officers to overcome barriers that preclude commercial firms from doing business with DOD. This training would be aimed primarily at using existing legal and regulatory latitude to procure products and services from the commercial sector. Training in critical thinking and problem-solving skills should be a prerequisite for the level III certification of acquisition personnel. Subsequent training should emphasize business acumen as well as critical thinking and problem solving.
- Use a range of resources to carry out this training in commercial practices, including private sector training, involvement in professional societies that include commercial firms, and government-industry personnel exchange. Internet technologies should be fully exploited for sharing lessons learned and best practices.
- Use the industry studies program of the Industrial College of the Armed Forces (ICAF) to broaden DOD's knowledge of the commercial industrial base. The key results of these studies should be made part of the curriculum and should be made available to policy makers and acquisition personnel.

Recommendation 3. Fund and execute additional rapid-response demonstration programs to build a broad base of experience with ICMM.

The DOD acquisition and sustainment management chain—USD(AT&L), service acquisition executives, center commanders, and PEOs/PMs—should encourage demonstration programs to build the base of experience needed for increased ICMM. This management chain should advocate, plan, program, budget, and protect the necessary resources for these demonstrations. It should

- Provide waivers, as needed, to experiment with commercial practices.
- Reexamine the present roles and missions of the Defense Production Act Titles I and III programs, including the Diminishing Manufacturing Sources and Industrial Base Analysis programs and processes, to ascertain their full potential for supporting a more robust transformation of the defense industrial base to an ICMM base.
- Focus and fund the DOD ManTech Program (and its service/DLA components) to exploit the full authority and latitude of ManTech's unique statutory mandate. ManTech should be aggressively employed to foster, execute, and transition significant demonstration projects with major ICMM implications to the DOD. Examples of specific projects include
  —Modeling-and-simulation-based ICMM "acquisition war games" undertaken in partnership with DAU, using authoritative experts as visiting professors and research fellows, as well as active or retired PMs and experts from industry and government. Provide early stimulus and support to the concept and formulation of a commercial acquisition academy;
  —Demonstration programs to fully explore and validate the optimum use of the commercial manufacturing base for military products;
  —Advanced ICMM technologies, including modeling and simulation tools and methods to faithfully represent the relationships between product design, commercial producibility, and life-cycle supportability;
  —Methods to achieve or approach six-sigma quality control on defense and commercial comanufacturing lines; and
  —Technologies and business and management practices that will enable affordable production of small lot sizes, the goal being an affordable lot size of one.

Recommendation 4. Contract for life-cycle support and technology refreshment.

For systems where commercially manufactured content is significant, DOD acquisition strategies should give total product responsibility to the system integrator for design, production, product support, and system availability, including responsibility for commercially based technology refreshment. The objective is to assure producible designs that can be supported with readily available components where underlying commercial technology in systems and subsystems is progressing rapidly. To accomplish this, the USD(AT&L) should issue the following implementing guidelines:

- Use only performance specifications.
- Give configuration management authority to the contractor to redesign and replace failed products so long as DOD retains plug-and-play (i.e., form, fit, and function) interchangeability.
- Provide incentives for contractors to initiate technology refreshment as required to support the product. Plan to minimize costs and time for requalification except where necessary for safety. Use interface control documents and functional specifications to manage the contractor.
- Use multiyear production pricing curves and award term contracts that motivate the design and production of reliable systems and cost-effective sustainment.
- Plan and fund functional product improvements separately from product support. However, with configuration management authority, contractors should be able to initiate replacement or upgrade of components when it is less costly than continuing to repair the existing ones.
- Demonstrate and encourage lean manufacturing and just-in-time supply-chain practices.

Recommendation 5. Increase DOD and defense sector awareness of planned and emerging commercial technologies and capabilities.

USD(A&T) and service acqusition executives (SAEs) should develop programs to increase the insight that DOD and defense contractors have into emerging commercial products and planned technologies, to assess capabilities that will be available in time frames of military interest, and to use this information in defining requirements for acquisition and upgrade programs. These new insight programs should include

- Designating and training DOD personnel who will specialize in commercial market research and advising the acquisition community on emerging commercial products and production capabilities that DOD can leverage.
- Supplementing this in-house capability with access to commercial market research companies and their knowledge bases.
- Encouraging defense prime contractors to establish business relationships that give them insight into the proprietary plans of commercial partners and allow them to assess ICMM possibilities for emerging commercial technologies and planned production capabilities.

Recommendation 6. Invest in R&D to increase the compatibility of military operating environments and commercially produced components.

The Director of Defense Research and Engineering should establish a program to develop generally applicable solutions to mitigate the effects of differences between military operating environments and commercial operating environments. Examples of solutions are designs and manufacturing methods that harden commercial items to perform reliably in military environments (temperature, humidity, shock, vibration, and radiation) and weapons system features that isolate or buffer commercial components from such environments. Experience from commercial industry in applying commercial components in severe environments should be analyzed to determine their applicability to military systems. Implementation should include

- Analyses to identify classes of commercial subsystems and components that offer the needed performance potential but are incompatible with military operating environments. These analyses should identify opportunities for generally applicable solutions that would have broad impact across new acquisitions and upgrades. Particular emphasis should be placed on commercial components and subsystems for multifunctional sensors, battlefield robots, missile seekers, night vision equipment, and embedded computers.
- Funding for R&D programs to develop the needed solutions. Since the objective is to increase the use of the commercial industrial base, this mission could be assigned to the DOD ManTech program.

DISCUSSION OF RECOMMENDATIONS

This section briefly discusses the committee's recommendations as they relate to the items in the statement of task.

Task 1. Identify advances in commercial technology and best practices that present opportunities for increased ICMM.

Based on a review of studies forecasting advances in manufacturing processes expected by 2010, particularly the *Integrated Manufacturing Technology Roadmapping Project* (IMTI, 2000), the committee's assessment is that the advances in technology and practices most important to increasing opportunities for ICMM are in the following areas:

- Product design, definition, and data interchange;
- Systems for manufacturing planning and execution;

- Collaborative design techniques, especially those that bring subcontractors into the process earlier;
- Enabling information infrastructure;
- Product modeling and simulation (including cost and producibility);
- Methods of assuring and improving quality;
- Test and evaluation;
- Design for product support and reduced life-cycle cost; and
- Integration planning and analysis.

To take full advantage of these advances, DOD needs to be more actively involved in commercial standards development processes to ensure early consideration of the interfaces needed for defense applications. Commercial industry will generally not make changes to accommodate defense needs once standards are set.

The most notable best practice is commercial industry's efforts to seek radically new manufacturing sources for components, driven by pressure to reduce production costs. A new industry, flexible electronic manufacturing services (EMS) companies, is available for dual production (see Finding 4). This trend toward flexible EMS companies appears to be continuing, along with trends in CAD, robotics, and other technologies. It is difficult to predict the consequences of these powerful forces in 2010 and beyond. However, it appears that two distinct lines of evolving commercial manufacturing will be available by 2010:

- Horizontal (outsourced) manufacturing by high-tech companies, where efficiency is achieved by specialized facilities operating in multiple shifts, and
- Vertical (in-house) manufacturing, where competitive costs are achieved by highly automated, flexible manufacturing that allows a wider spectrum of assembly than specialized horizontal manufacturing.

The committee believes that both capabilities will be available to the military once the barriers pointed out in this report have been dismantled.

Task 2. Identify technology areas where rapid commercial advances could be leveraged.

The trend toward increased miniaturization and integration of MOEMS will, by 2010, offer high-capital-investment facilities capable of making defense as well as commercial subsystems (see Finding 4).

Task 3. Identify major weapon systems suitable for production by commercial enterprises.

The committee finds that major weapon systems in 2010 and beyond will continue to be integrated, assembled into final products, and tested by defense prime contractors. However, virtually *all* defense systems have many opportunities for subsystems and components to be bought from commercial suppliers, either as COTS insertions or as custom-produced items (see Finding 4). While progress has been made with COTS end items and components, the opportunities for custom items using commercial production capabilities have not been well capitalized on. There is also a broad range of commercial practices that can be applied throughout the life cycle of virtually all military systems.

Task 4. Identify barriers limiting ICMM.

The committee finds that barriers were well defined in earlier studies and that most of the needed tools are available. Further implementation action is needed, and as described in Chapter 3, there is now a body of experience to build on.

8

References

Detlefs, B. 1998. "MEMS 1998—Emerging Applications and Markets," *mst News* (4) 27. Available online at <http://www.vdivde-it.de/mstnews/pdf_aktuell/news0498.pdf>. Accessed on September 6, 2001.

DOD (Department of Defense). 1989. *Report of the Defense Science Board on Use of Commercial Components in Military Equipment*, June. Washington, D.C.

DOD. 1993. *Report of the Defense Science Board on Defense Manufacturing Enterprise Strategy*, September. Washington, D.C.

DOD. 2000a. *Report of the Defense Science Board on the Technology Capabilities of Non-DoD Providers*, June. Washington, D.C.

DOD. 2000b. *Report of the Defense Science Board on Preserving a Healthy and Competitive U.S. Defense Industry to Ensure Our Future National Security*, November. Washington, D.C.

Gansler, J. 1995. *Defense Conversion: Transforming the Arsenal of Democracy*. MIT Press, Cambridge, Mass.

GAO (General Accounting Office). 1998. *Best Practices: Successful Application to Weapon Acquisitions Requires Changes in DOD's Environment*, February 24 (GAO/NSIAD-98-56).

Global Information, Inc. Undated. "Military Leverages Commercial Opportunities," *Microengineering and Nanotechnology News* (Kawasaki, Japan). Available online at <http://www.gii.co.jp/japanese/bc4376_nl_microtechnology_sample.html>. Accessed on October 3, 2001.

IMTI (Integrated Manufacturing Technology Initiative). 2000. *Integrated Manufacturing Technology Roadmapping Project: An Overview of the IMTR Roadmaps*. IMTI, Oak Ridge, Tenn.

Lee, D. 2001. Testimony of the Director, Defense Procurement, Office of the Under Secretary of Defense (Acquisition, Technology & Logistics) before the U.S. House of Representatives Committee on Government Reform, Subcommittee on Technology and Procurement Policy, July 17. Available online at <http://www.house.gov/reform/tapps/hearings/7-17-01/Lee.htm>.

Lockheed Martin. 2001. "Lockheed Martin, Fleet Ballistic Missile Open System Architecture Team Earns DOD Acquisition Reform Award." Press release, Lockheed Martin Space Systems Company, Sunnyvale, Calif. Available online at <http://biz.yahoo.com/bw/011002/21019_1.html>. Accessed on October 22, 2001.

Miscioscia, L., and A. Libin. 2000. *It's a Tech Revolution and EMS Companies Are Building It: A Growth Story*. Lehman Brothers, New York.

NRC (National Research Council). 1993. *Science, Technology, and the Federal Government: National Goals for a New Era*. National Academy Press, Washington, D.C.

NRC. 1999a. *Defense Manufacturing in 2010 and Beyond.* National Academy Press, Washington, D.C.

NRC. 1999b. *U.S. Industry in 2000: Studies in Competitive Performance*. National Academy Press, Washington, D.C.

NRC. 2001. *Trends in Federal Support of Research and Graduate Education,* p. 132 (prepublication copy). National Academy Press, Washington, D.C.

OTA (Office of Technology Assessment). 1994. *Assessing the Potential for Civil-Military Integration: Technologies, Processes, and Practices*. September 1994, OTA-ISS-611.

OTA. 1995a. *Assessing the Potential for Civil-Military Integration: Selected Case Studies*. September 1995, OTA-BP-ISS-158.

OTA. 1995b. *Other Approaches to Civil-Military Integration: The Chinese and Japanese Arms Industries*. March 1995, OTA-BP-ISS-143.

SAIC (Science Applications International Corporation). 1999. *Innovation in Contractual Incentives, Phase 1 Report*, Appendix D. Prepared for the Deputy Assistant Secretary of the Army (Procurement). Available online at <http://acqnet.sarda.army.mil/library/final/appenD.htm.> Accessed on May 10, 2001.

Tang, W.C. 1999. MEMS at DARPA, presentation, December 2. Available online at <http://www.glennan.org/memsdarpa.pdf>. Accessed on December 28, 2001.

USD(AT&L) (Under Secretary of Defense (Acquisition, Technology & Logistics)). 2000. *The Acquisition 2005 Task Force Final Report: Shaping the Civilian Acquisition Workforce of the Future,* October. USD(AT&L) and Under Secretary of Defense (Personnel and Readiness), Washington, D.C.

USD(AT&L). 2001. *Intellectual Property: Navigating Through Commercial Waters—Issues and Solutions When Negotiating Intellectual Property with Commercial Companies*, Version 1.1, October 15.

Winner, R., and L. Griffin. 1998. *Civil-Military Industrial Integration: Dual Production of Military Electronics* (rev. March 1999). R. Winner & Associates, Hopkinton, Mass.

Appendices

A

Biographical Sketches of Committee Members

Michael F. McGrath, *Chair,* is vice president for government business at the Sarnoff Corporation, a leading R&D company with both commercial and government clients. He is responsible for program development across all Sarnoff business units to meet government needs for innovative, dual-use technologies in networks and information systems, sensors and microelectronics, and biotechnology. Dr. McGrath holds a B.S. in space science and applied physics, an M.S. in aerospace engineering both from Catholic University, and a doctorate in operations research from the George Washington University. Prior to joining Sarnoff, Dr. McGrath was the assistant deputy under secretary of defense (dual use and commercial programs), where he was responsible for policy, strategies, and programs to help the services and defense agencies make routine use of commercial technologies and industrial capabilities in defense systems. He directed the Commercial Technology Insertion Program, the Commercial Operating and Support Savings Initiative, and DOD's Title III investment program to establish sources for critical materials and technologies. Dr. McGrath also served as the assistant director for manufacturing in DARPA's Defense Sciences Office. He was the agency focal point for manufacturing technology and the program manager for three DARPA programs (Agile Manufacturing, Electronic Commerce Resource Centers, and Affordable Multi-Missile Manufacturing). He also served in leadership positions for several DOD-wide initiatives to improve manufacturing and reduce the cost of defense systems. Before that, Dr. McGrath was the director of the DOD CALS Office in the Pentagon, where he guided the Computer-Aided Acquisition and Logistics Support program from its inception. He has also held positions in weapon system logistics planning and management, first at the Naval Air Systems Command and later in the Office of the Secretary of Defense.

Walter B. LaBerge (NAE), *Vice Chair,* is currently visiting professor at the Naval Postgraduate School, Monterey, California, and the Institute for Advanced Technology at the University of Texas, Austin. He received his Ph.D. in physics from the University of Notre Dame. He has spent half a career in the government as a past chairman of the Army Science Board, under secretary of the army, acting director of defense (research and engineering), assistant secretary of USAF (R&D), technical director of NWC China Lake, and assistant secretary-general of NATO. The other half of his career was with the aerospace industry in executive positions with Philco-Ford and the Lockheed Corporation.

Arden L. Bement, Jr. (NAE), is the Basil S. Turner Distinguished Professor of Engineering at Purdue University, where he has been on the faculty since 1993. His expertise in materials science and nuclear engineering includes experience in irradiation effects, deformation and fracture, energy materials technology, and defense materials technology. His most recent research has been in electroceramics, especially high-temperature superconductors and ferroelectric ceramics. He continues his interest in nuclear materials management, including nuclear waste management and plutonium management for nonproliferation of nuclear weapons. Before Purdue, Bement was vice president for science and technology and vice president for technical resources at TRW, Inc. He was Deputy Under Secretary of Defense (Research and Engineering) and director of the materials science office at DARPA. He has held positions at MIT, Battelle Northwest Laboratories, and Hanford Laboratories (General Electric). He is a fellow of the American Society for Metals International, the American Nuclear Society, and the American Institute of Chemists, and is a member of Sigma Xi. Dr. Bement is a member of the National Academy of Engineering and an honorary member of the American Ceramics Society.

Peter DeMayo was the vice president for contract policy at Lockheed Martin until he retired in March 2000 and is now pursuing interests relating to education and defense acquisition. He served in the United States Navy Supply Corps with primary duty in the acquisition of major systems. Mr. DeMayo was assistant professor of system acquisition management at the Naval Postgraduate School and commander of the Naval Aviation Depots, a logistics support organization of six organic depots, 25,000 people, and sales of $2 billion per year. He retired from the Navy in February 1990 with the rank of Rear Admiral. Mr. DeMayo is a member of the Board of Visitors of the Defense Acquisition University and the Defense Systems Management College, the Board of Directors of the Procurement Roundtable, and the Board of Advisors at the George Washington University School of Business and Public Management.

Gary Denman is president and CEO for GRC International and AT&T Company. Prior to joining GRC in March 1995, he served as deputy director and director of the DOD's Advanced Research Projects Agency (ARPA) for 5 years. Prior to that, he served within the Air Force's Wright Laboratories in various capacities that included director of the Air Force's Manufacturing Technology Program and director of the Air Force Materials Laboratory. He has been on the forefront of advanced manufacturing technology, including computer-integrated manufacturing and manufacturing processes for electronic components, metals, and composites. While director of ARPA, he was a project sponsor at the NRC, where he forcefully emphasized the importance of manufacturing activities.

Joseph A. Heim is a member of the corporate materials engineering staff at Genie Industries, an international manufacturer of man-lift and materials handling equipment. He earned a Ph.D. in industrial engineering from Purdue University and a B.S. in mechanical engineering and an M.E. in computer science from the University of Louisville. Prior to joining Genie Industries, Dr. Heim was a member of the industrial engineering faculty at the University of Washington in Seattle and held various executive positions at InfoSystems Design and Integrated Production Systems. From 1990 to 1992 he was the J. Herbert Hollomon Fellow at the National Academy of Engineering.

F. Suzanne Jenniches is the vice president of Northrop Grumman ESSS Communication Systems with operating units in Baltimore, Maryland; Cincinnati, Ohio; Market Deeping, in England; and Oslo, in Norway. Prior responsibilities included nondefense automation and information systems, B-1B offensive radar operations, robotics, and defense electronics manufacturing and testing. She serves as a consultant on the U.S. Army Science Board. She is also a past president of the Society of Women Engineers.

James Mattice is director of management and organizational development at Universal Technologies Corporation. He is responsible for ongoing government and commercial activities in research, development, and technology transition. Previously, Mr. Mattice served as Air Force executive in residence at the Federal Executive Institute in Charlottesville, Virginia; deputy assistant secretary of the Air Force for research and engineering in the Office of the Secretary of the Air Force, the Pentagon; executive director in the Office of the Commander, Director of Development Planning; and a variety of senior management positions in Air Force laboratories at the Aeronautical Systems Center, Wright-Patterson Air Force Base, Ohio. These Air Force positions capped 38 years of experience with in-house laboratory research, as well as leadership in all aspects of basic research, exploratory, advanced development, manufacturing technology, and executive development programs and organizations. Mr. Mattice is internationally recognized for his accomplishments as a research and development organization leader, a corporate strategic planner, an international cooperative program architect, and an agent for interservice, interagency collaborative policy and programs. He has served on numerous boards, special study panels, and advisory committees in government and with industry and academia, here and abroad.

David R. Smith is director of the Production Systems Engineering and Technology Organization at Eastman Kodak Research and Development. He began his career with Eastman Kodak in 1965 and has worked in design engineering, technology development, process improvement, and systems engineering. A native of Massachusetts, Mr. Smith has an M.S. in electrical engineering from the University of Rochester and a B.S. in electrical engineering from the University of Massachusetts. He is a member of the Board of Directors for Microelectronics & Computer Technology Corporation and the Optoelectronics Industry Development Association.

Robert I. Winner is the president of R. Winner & Associates, Hopkinton, Massachusetts, a consulting and facilitation firm focusing on integrated development processes, defense acquisition reform, and information technology planning and implementation. Dr. Winner earned a Ph.D. in information and computer science from the Georgia Institute of Technology, an M.S. in computer science from Purdue University, and a B.S. in mathematics from Union College. He is a graduate of Harvard University's John F. Kennedy School of Government Senior Executive Program in National and International Security and the ITT program on design of industrial experiments. Dr. Winner previously was vice president and secretary of the Enterprise Computing Institute, the principal scientist of the Center for High Performance Computing, deputy director of the Computer and Software Engineering Division of the Institute for Defense Analyses, and associate professor of computer science at Vanderbilt University. He was the principal author and team leader of the landmark 1988 IDA study *The Role of Concurrent Engineering in Weapons System Acquisition*, which defined concurrent engineering and introduced the idea of integrated product and process development into the defense community, and he was chief of staff of two follow-on Defense Science Board studies. In 1999 he published an in-depth, case-based study, *Civil-Military Industrial Integration: Dual Production of Military Electronics*, which proposed and analyzed new approaches to the manufacture of DOD electronics. He has been a member of the Education Board of the Association for Computing Machinery and is recipient of the ACM's Outstanding Service Award.

B

Briefings

The following individuals made presentations to the Committee on Integration of Commercial and Military Manufacturing in 2010 and Beyond:

1. Overview of ManTech Program and Sponsor's Goals, Steve Linder, *Office of Naval Research*

2. Dual Production of Military Electronics, Larry Griffin, *independent consultant*

3. Acquisition Reform Initiatives, Stan Soloway, *Deputy Under Secretary of Defense, Acquisition Reform*

4. Military Products from Commercial Lines, Brench Boden, *Wright-Patterson AFB*

5. Military and Commercial Turbine Engines, Larry Trowel, *General Electric Aircraft Engines*

6. A Common Manufacturing Base for Dual-Use Applications, Jim Fallon and Joe Thomas, *M/A-COM*

7. Miniature Air-Launch Decoy (MALD) Program, Lt. Col. Walter Price, *U.S. Air Force, Wright-Patterson AFB*

8. Circuit Board Manufacturing at SCI Systems, Steve Werner and Jerry Thomas, *SCI Systems*

9. Dual Production Experience at Rockwell-Collins, Herm Reininga, *Rockwell-Collins*

10. Army Experience and Opportunities for Commercial Leverage in Army Transformation Programs, LTG Paul Kern, *U.S. Army*

11. Experience with Implementation of CMI in the Defense Department, Jacques Gansler, *University of Maryland*

12. Reduced Total Cost of Ownership (RTOC) Pilot Projects in DOD, Spiros Pallas, *Principal Deputy Director, Strategic and Tactical Systems, Office of the Secretary of Defense*

13. Commercial and Military Interaction in U.S. Army Night Vision Technology, Fenner Milton, *U.S. Army Night Vision Center*

C

Background and History

BACKGROUND

Integrating commercial and military manufacturing is not a new idea. Mobilization of the U.S. commercial industrial base into the "arsenal of democracy" was essential to victory in World War II.

After the war, this capacity reverted to its civilian uses. Through the 1950s, the military became a major driver of technological change in the American economy, spawning industries like commercial aeronautics, semiconductors, and communications satellites (NRC, 1993).

The Cold War era saw the development of a permanent defense industrial base and defense acquisition practices distinctly different from commercial practice. The U.S. military's demand for weapons continued to be strong until the end of the Cold War. The permanent armaments industry met the requirements for military acquisition, despite periodic unease with the cost of weapons and accusations of waste, fraud, and abuse. A complex business relationship evolved between government and contractors; laws and regulations governing the procurement process became increasingly complicated.

The personnel required for contractors to comply with these regulations became a growing cost burden to companies and their government business units (SAIC, 1999). One reaction to the regulations was the separation for accounting purposes of commercial and government business in companies that do both (Gansler, 1995). This separation shields commercial businesses from DOD audits and the possibility of prosecution for irregularities. Auditability has become a first-order driver of defense contractor development and production processes.

History of Acquisition Reform

Since 1990, a number of trends have combined to begin the process of change in DOD's acquisition practices, including debate over the vision of civil-military integration (CMI). Box D-1 provides a timeline of acquisition reform initiatives in recent years.

The first of these trends was reduced budgets for the acquisition of new defense systems, resulting largely from the end of the Cold War. Simultaneously, the capabilities of commercial equipment and its application for military use have

Timeline of Acquisition Reform Since 1994

1994 Secretary Perry's Mandate for Change

- Rapidly acquire commercial products and technology.
- Introduce reforms to minimize the use of military specifications.
- Assist in conversion of U.S. defense-unique companies to dual-use (producing for both commercial and military) production.
- Aid in transfer of military technology to the commercial sector.
- Preserve defense-unique core capabilities.
- Adopt business practices characteristic of world-class customers and suppliers and encourage DOD suppliers to do the same.

1994 Federal Acquisition Streamlining Act

- Permitted purchases of items below $2,500 without competitive quotes or compliance with Buy American Act and certain small business requirements.
- Exempted commercial product procurements from laws, including submission of cost or pricing data and cost accounting standards (CAS) requirements; established preference for commercial items.

1996 Federal Acquisition Reform Act and Information Technology Management Reform Act (Collectively Known As the Clinger-Cohen Act)

- Required that agency heads establish a process to select, manage, and control their IT investments.
- Authorized a test of the use of the Simplified Acquisition Procedures (SAP) for commercial items between the simplified acquisition threshold of $100,000 and $5 million.

1997 Rewrite of DOD 5000 Series of Acquisition Regulations

- Incorporated new laws and policies in a dramatically shorter document.

NOTE: Adapted from <www.dsmc.dsm.mil/jdam/contents/legislation.htm>.

Box D-1

accelerated, especially in electronics. These trends, as well as changes in the nation's defense strategy, have caused uncertainty over which weapons to buy and interest in broadening the existing stand-alone defense industrial base.

SUMMARY OF EARLIER STUDIES

Since the late 1980s, studies have discussed commercial-military integration, also known as civil-military integration, and described its anticipated benefits (both terms are known by the acronym CMI.[1]) The studies were prepared by organizations such as the Defense Science Board (DOD, 1989), the Office of Technology Assessment (OTA, 1994, 1995a, and 1995b) and the General Accounting Office (GAO, 1998).

The DSB study, which focused on semiconductor devices, noted that "despite overwhelming verbal support, movement towards greater use of commercial products and practices has been slow." The transmittal letter included in that report recommended the use of open systems architecture, which would allow increased use of commercial hardware and software and the corresponding industrial base while demonstrating the ability to use commercial subsystems and nongovernment standards as well as their benefits. It also recommended the use of demonstration programs to test the ability to buy commercial products and services using commercial practices.

The 1994 OTA study *Assessing the Potential for Civil-Military Integration: Technologies, Processes, and Practices* noted that Congress had long been interested in CMI. This interest was evident as early as 1984, when the Defense Procurement Reform Act mandated that DOD use "standard or commercial parts" when developing or buying military-unique products whenever technically or economically feasible. The 1994 OTA study also noted that the 1990, 1991, and 1993 Defense Authorization acts all contained language promoting CMI.

Based on analysis of previous studies, the 1994 study suggested that military and commercial manufacturing had become segregated for a number of reasons:

- Acquisition laws, regulations, and culture;
- Military specifications and standards;
- Militarily unique technologies or products;
- Commercially uneconomical orders;
- Emphasis on performance rather than costs; and
- Classified technologies.

---

[1]The committee chose to use the acronym ICMM, for integration of commercial and military manufacturing, rather than the acronym CMI, for civil-military integration or commercial-military integration. The committee believes that ICMM emphasizes manufacturing and the optimum use of the commercial industrial base. However, this appendix uses the acronym CMI where it was used by earlier studies.

It also noted that the structure of acquisition laws and regulations and the culture resulting from them provide few incentives to integrate. OTA also found, however, that technological developments had reinforced the trend toward integration of the commercial and military technological and industrial bases. While it noted that some case studies cited the potential for cost savings of 20 to 60 percent, it believed that savings were difficult to quantify and that it was unlikely that such savings were possible across the entire industrial base. Finally, OTA believed that the most important contribution of CMI might not be in savings, but rather in preserving the ability to support national security objectives in the face of spending reductions.

The OTA (1995a) study *Assessing the Potential for Civil-Military Integration: Selected Case Studies* stated as follows: "Despite several previous initiatives to promote integration, much of the DTIB [Defense Technology and Industrial Base] remains isolated. Still, significant CMI [Civil-Military Integration] currently exists." OTA noted that it believed CMI was more likely at the lower industrial tiers than at the prime contractor level, where components are combined to produce military-unique systems. OTA stated that technology trends had blurred differences between commercial and military technology in its three case studies: flat-panel displays, polymeric composites, and shipbuilding. Finally, this OTA study concluded that both increased access to commercial technology and reduced system cost were likely results of CMI, but that both benefits were difficult to quantify.

Sources outside government also recognize the potential of ICMM. The 1999 NRC study *Defense Manufacturing in 2010 and Beyond: Meeting the Changing Needs of National Defense* focused on leveraging advances in commercial manufacturing in its recommendations. The study identified the challenges of using commercial manufacturing capacity, including the use of and design for commercial processes, the incorporation of COTS parts and subsystems into military equipment, reforms in acquisition procedures to accommodate commercial practices, and incentives for commercial industry to manufacture defense parts. It identified nontechnical barriers, such as accounting practices and acquisition regulations, as the most serious impediment to achieving this challenge. The current study pursues these challenges in greater depth.

Given the limited progress in CMI despite the long interest in the area, Gansler (1995) states that "perhaps no point . . . should be more heavily emphasized than the extreme difficulty that will be faced in attempting to implement the changes required for civil/military integration." He identified the primary causes of this difficulty as (1) the complex nature of overcoming the status quo and the investment of careers in working within the current system and (2) the differences between commercial and military procurement systems brought about by an inherent distrust of defense contractor motives.

The 1993 Defense Science Board study (DOD, 1993) addressed that difficulty in attempting to implement change in defense acquisition. It emphasizes that what needs to be done is well documented, and that the focus should be on implementation methods. The present NRC study took a similar approach, focusing on implementation.

D

Acronyms and Abbreviations

ACAT	acquisition category
ASICs	application-specific integrated circuits
CAD/CAM	computer-aided design/computer-aided manufacturing
CAIV	cost as an independent variable
CAS	cost accounting standards
CMI	civil-military integration or commercial-military integration
COTS	commercial off-the-shelf
DARPA	Defense Advanced Research Projects Agency
DAU	Defense Acquisition University
DFARS	Department of Defense FAR Supplement
DOD	Department of Defense
DSMC	Defense Systems Management College
DTIB	defense technology and industrial base
EMS	electronic manufacturing services
FAR	Federal Acquisition Regulations
FBM	fleet ballistic missile
GaAs	gallium arsenide
GAO	General Accounting Office
GEAE	General Electric Aircraft Engines
GOTS	government off-the-shelf

GPS	global positioning system
IC	integrated circuit
ICAF	Industrial College of the Armed Forces
ICMM	integration of commercial and military manufacturing
IMTI	integrated manufacturing technology initiative
IP	intellectual property
IPPD	integrated product and process development
IPT	integrated product team
IR	infrared
MALD	miniature air-launched decoy
ManTech	DOD Manufacturing Technology Program
MEMS	microelectromechanical systems
MIL-STD	military standard
MOEMS	micro-opto-electromechanical systems
MOSA	modular open systems architecture
MPCL	military products from commercial lines
NC	numerically controlled
NRC	National Research Council
OEMs	original equipment manufacturers
OTA	Office of Technology Assessment
PEM	plastic encapsulated microcircuit
PEOs	program executive officers
PLGR	precision lightweight GPS receiver
PNP	pulse narrow-band preprocessor
PWA	printed wiring assembly
R&D	research and development
RDT&E	research, development, test, and evaluation
RFP	request for proposal
RTOC	reduced total cost of ownership
SAEs	service acquisition executives
SAP	Simplified Acquisition Procedures
USAF ESC	U.S. Air Force Electronic Systems Center
USD(AT&L)	Undersecretary of Defense (Acquisition, Technology and Logistics)
USN	U.S. Navy

E

ManTech Statute

UNITED STATES CODE
TITLE 10—ARMED FORCES
Subtitle A—General Military Law
PART IV—SERVICE, SUPPLY, AND PROCUREMENT
CHAPTER 148-NATIONAL DEFENSE TECHNOLOGY AND INDUSTRIAL BASE, DEFENSE REINVESTMENT, AND DEFENSE CONVERSION

SUBCHAPTER IV—MANUFACTURING TECHNOLOGY

§ Sec. 2521. - Manufacturing Technology Program

(a) Establishment.—

The Secretary of Defense shall establish a Manufacturing Technology Program to further the national security objectives of section 2501(a) of this title through the development and application of advanced manufacturing technologies and processes that will reduce the acquisition and supportability costs of defense weapon systems and reduce manufacturing and repair cycle times across the life cycles of such systems. The Secretary shall use the joint planning process of the directors of the Department of Defense laboratories in establishing the program. The Under Secretary of Defense for Acquisition, Technology, and Logistics shall administer the program.

(b) Purpose of Program. —

The Secretary of Defense shall use the program -

(1) to provide centralized guidance and direction (including goals, milestones, and priorities) to the military departments and the Defense Agencies on all matters relating to manufacturing technology;

(2) to direct the development and implementation of Department of Defense plans, programs, projects, activities, and policies that promote the development and application of advanced technologies to manufacturing processes, tools, and equipment;

(3) to improve the manufacturing quality, productivity, technology, and practices of businesses and workers providing goods and services to the Department of Defense;

(4) to focus Department of Defense support for the development and application of advanced manufacturing technologies and processes for use to meet manufacturing requirements that are essential to the national defense, as well as for repair and remanufacturing in support of the operations of systems commands, depots, air logistics centers, and shipyards;

(5) to disseminate information concerning improved manufacturing improvement concepts, including information on such matters as best manufacturing practices, product data exchange specifications, computer-aided acquisition and logistics support, and rapid acquisition of manufactured parts;

(6) to sustain and enhance the skills and capabilities of the manufacturing work force;

(7) to promote high-performance work systems (with development and dissemination of production technologies that build upon the skills and capabilities of the work force), high levels of worker education and training; and

(8) to ensure appropriate coordination between the manufacturing technology programs and industrial preparedness programs of the Department of Defense and similar programs undertaken by other departments and agencies of the Federal Government or by the private sector.

(c) Execution. —

(1) The Secretary may carry out projects under the program through the Secretaries of the military departments and the heads of the Defense Agencies.

(2) In the establishment and review of requirements for an advanced manufacturing technology or process, the Secretary shall ensure the participation of those prospective technology users that are expected to be the users of that technology or process.

(3) The Secretary shall ensure that each project under the program for the development of an advanced manufacturing technology or process includes an implementation plan for the transition of that technology or process to the prospective technology users that will be the users of that technology or process.

(4) In the periodic review of a project under the program, the Secretary shall ensure participation by those prospective technology users that are the expected users for the technology or process being developed under the project.

(5) In order to promote increased dissemination and use of manufacturing technology throughout the national defense technology and industrial base, the Secretary shall seek, to the maximum extent practicable, the participation of manufacturers of manufacturing equipment in the projects under the program.

(6) In this subsection, the term ''prospective technology users'' means the following officials and elements of the Department of Defense:

(A) Program and project managers for defense weapon systems.
(B) Systems commands.
(C) Depots.
(D) Air logistics centers.
(E) Shipyards.

(d) Competition and Cost Sharing. —

(1) In accordance with the policy stated in section 2374 of this title, competitive procedures shall be used for awarding all grants and entering into all contracts, cooperative agreements, and other transactions under the program.

(2) Under the competitive procedures used, the factors to be considered in the evaluation of each proposed grant, contract, cooperative agreement, or other transaction for a project under the program shall include the extent to which that proposed transaction provides for the proposed recipient to share in the cost of the project. For a project for which the Government receives an offer from only one offeror, the contracting officer shall negotiate the ratio of contract recipient cost to Government cost that represents the best value to the Government.

(e) Five-Year Plan. —

(1) The Secretary of Defense shall prepare a five-year plan for the program which establishes —

(A) the overall manufacturing technology goals, milestones, priorities, and investment strategy for the program; and

(B) for each of the five fiscal years covered by the plan, the objectives of, and funding for the program by, each military department and each Defense Agency participating in the program.

(2) The plan shall include the following:

(A) An assessment of the effectiveness of the program, including a description of all completed projects and status of implementation.

(B) An assessment of the extent to which the costs of projects are being shared by the following:

(i) Commercial enterprises in the private sector.

(ii) Department of Defense program offices, including weapon system program offices.

(iii) Departments and agencies of the Federal Government outside the Department of Defense.

(iv) Institutions of higher education.

(v) Other institutions not operated for profit.

(vi) Other sources.

(C) Plans for the implementation of the advanced manufacturing technologies and processes being developed under the program.

(3) The plan shall be updated annually and shall be included in the budget justification documents submitted in support of the budget of the Department of Defense for a fiscal year (as included in the budget of the President submitted to Congress under section 1105 of title 31).